DIRECT PROTECTION OF INNOVATION

Report of a study prepared under contract for the Technological Information and Patents Division of the Commission of the European Communities, Directorate-General Telecommunications, Information Industries and Innovation

Direct Protection of Innovation

edited by

William Kingston
Trinity College, University of Dublin,
Ireland

Springer-Science+Business Media, B.V. 1987

Library of Congress Cataloging in Publication Data

Direct protection of innovation.

Includes index.
1. Patent laws and legislation--European Economic
Community countries. 2. Technological innovations--
Law and legislation--European Economic Community
countries. I. Kingston, William.
KJE2725.D57 1987   341.7'586'094   86-27868

ISBN 978-94-017-1267-5   ISBN 978-94-017-1265-1 (eBook)
DOI 10.1007/978-94-017-1265-1

EUR 10451 EN

Book information

Publication arranged by: Commission of the European Communities, Directorate-General Telecommunications, Information Industries and Innovation, Luxembourg

Copyright/legal notice

TABLE OF CONTENTS

FOREWORD

1 When in the late seventies the Commission started its preparatory work on a plan of action(1) for the creation of a real Community Innovation market, obviously the question came up, how patents or industrial property could specifically help in stimulating technological and industrial innovation and technology transfer in the Community. From an earlier contractor's study(2) about possible items of action in patent law for the improvement of the impact of patents and patenting upon innovation, it was clear that, in principle, there was room for improvement but shortage of information as to how and to what extent efficient improvements should and could be made.

2 In the early 1980s then, the need for clarifying the potential for improvement in patent law and patent practice became more pressing, so that the Commission convened an informal meeting of experts on 3/4 November 1982 in Luxembourg, in order to discuss the issues relevant to the relationship between patent protection and innovation and to identify suitable subject matter for action or study. 38 experts from nine Member States, coming from different areas of activity in industrial property or in innovation attended the meeting, which was chaired by two of them.

3 From the papers presented and the discussion that followed it became clear that the issues were very complex and difficult and that, when reflecting about "improvements", a "reform", even radical, of the patent law or concept, could not be excluded from such reflexions. A list of about 30 topics for action or study by the Commission was drawn up, ending with the following final considerations:–

> "Patents play a central role in the innovation process, which extends from the scientific and technical research stage to the manufacture and marketing of new products and processes.
>
> In consequence of that, the experts invited to the meeting by the Commission of the European Communities considered that the task ahead lay in revitalising the patent system and making it more effective. The relationship between patents and industrial innovation should also be better organized, thereby improving the opportunities open to those who launch new products and processes on to the market".

4 In pursuance of the proposals listed, the Commission chose a number of the most important for further study. Five of these referred to improvements of the classical patent system, and one involved new concepts for the protection of innovation.

The "classical" studies referred to are:

(a) The period of grace
(b) The inventive step
(c) The cost of patents
(d) The patent litigation problem
(e) The term of the patent.

All of these studies were finished by the end of 1985 and are available in copy format, except study (b) which had to be dropped provisionally.

The "new concepts" study referred to "Direct protection of innovation", which is the object of this report in book format.

5 Trying to devise a new concept of protection for innovation is a venture, like innovation itself. But it is worth trying, in order to explore the potential of action that is available.

It is quite clear that we wish and need to build upon the existing systems of protection. But that should not prevent us from "inventing around" and even "innovating", if our researchers, inventors, innovators and industrialists should ask for it.

So, the purpose of publishing this report is essentially to stimulate the debate. It should be read for what it is : the product of original intellectual effort and research. It does however in no sense represent the opinions of the Commission of the European Communities.

R.K. APPLEYARD
Director-General, Information, Market and Innovation
Commission of the European Communities

(1) Later became:
"Council Decision of 25 November 1983 concerning a plan for the transnational development of the supporting infrastructure for innovation and technology transfer (1983–1985)", Official Journal of the European Communities, L 353, 15 December 1983, (viz. esp. action 2.2 (a), chap. 2, Annex I).

(2) "Patentschutz und Innovation", 1973, Max-Planck Institut für internationales und auslandisches Patent-, Urheber- und Wettbewerbsrecht, Munchen.

EDITOR'S PREFACE

This work was carried out under European Economic Community Study Contract No. PAT 1-83. It is structured according to the method used in the enquiry and reflection: thesis, criticism and response.

Two chapters, each dealing with a specific proposal for protecting innovation directly, are preceded by a chapter on the need for such protection and followed by another chapter which sets out the advantages which would result from adoption of such a system.

These four chapters, for which I am responsible, were then submitted for critical assessment to the following experts with reputations for study and understanding of Patents:–

ANDRE PIATIER, the author of many works on innovation, including a recent study of Barriers to Innovation, for the European Commission.

THOMAS MANDEVILLE AND STUART MACDONALD, both of the University of Queensland, Australia. They have carried out several empirical studies of Patent performance, and contributed to the recent Australian Government enquiry into the Patent system.

GORDON TULLOCK, of the Center for Study of Public Choice in Virginia, has written many books on public policy issues, and was commissioned by the United States Patent Office to examine and comment upon its operations.

AUBREY SILBERSTON, Professor at Imperial College, University of London, is joint author of the definitive study of the British Patent system.

HENK WOUTER DE JONG, Professor of Economics in the University of Amsterdam, who has made the economics of industrial innovation one of his special fields of study.

BRIAN WRIGHT, Associate Professor of Resource Economics at the University of California, Berkeley. He is the leading figure in the "new economics of Patents", on which he has published several important papers.

ANDRE BOUJU, a European Patent Attorney who also teaches in the Centre d'Etudes Internationales de la Proprieté Industrielle of the University of Strasbourg.

Chapters V to XI therefore contain the observations of these authors on both the proposals and their supporting arguments. Responses to these 'critical' Chapters by both authors of the proposals for Direct Innovation Protection complete the study.

Although they were unable to respond to an invitation to contribute chapters, certain other experts of international standing gave significant assistance to the project, especially to the final 'response' chapter. This was in the form of papers as well as lengthy discussions, and it is gratefully acknowledged. These experts were:–

Professor O.J. Firestone, whose work led to Canada investigating formally whether it should continue with a Patent system or not;

Professor Richard R. Nelson of Yale, whose writings on innovation policies form part of the basic corpus of learning in the field; and

Professor Morton I. Kamien, of Northwestern University, who is the author of several important works on the relationship between industry structure and innovation.

Finally, special thanks are due to Dr. H. Kronz, who has been the moving spirit in all the studies listed by Sir Raymond Appleyard. In the case of the present work, not alone is he the originator of one of the proposals investigated, he has also been an unfailing source of help and encouragement throughout.

Miss Wyn Sheerin handled the production of several difficult drafts of the 'Thesis' and 'Response' Chapters and the final processing of the complete text of the book, with exceptional devotion and competence.

WILLIAM KINGSTON,
John Good Senior Lecturer (Innovation),
Trinity College,
Dublin.

SUMMARY OF CHAPTERS I–IV
(THE 'THESIS' CHAPTERS)

Chapter I calls attention to the way in which Patents originated as grants of monopoly for actually "performing" something new. It then traces how and why they came to be given instead for "teaching" the novelty. It is the invention that is now protected directly, so that whatever protection investment in innovation receives, depends upon how strong the invention-innovation link is. In pharmaceuticals, the link is strong, so even indirect protection is valuable; in other areas it is so tenuous that the protection is worth little or nothing. Thus, only a fraction of the potential of the Principle of Patenting – the grant of monopoly in exchange for an economic activity that is considered to be socially desirable – has so far been exploited. Declining statistics of national usage of the classical Patent system signal the need for change.

Chapters II and III describe two proposals for giving protection to innovation directly, the "Innovation Patent" and the "Innovation Warrant". Both agree that the subject-matter of protection should be innovation, not invention; that any economic object can be protected, not just technology; that the term of grant should be variable, as a practicable surrogate for using money as the measure of monopoly; that grants should be incontestable unless obtained by fraud; that terms of grant may differ between regions of a country; and on several procedural aspects.

They differ in four main ways. The Innovation Patent requires that the "Innovation Object" must exist before protection is given, but the Warrant offers protection earlier, for investment to produce it. The "variable term" arrangements of the Patent strive to match the innovative capacity of the individual Patentee exactly, but the Warrant system uses "categories" of risk, so as to minimise administrative discretion. Such discretion is also involved in the licensing provisions of the Innovation Patent, but absent from those of the Warrant. In contrast, protection of the Innovation Patent grant through the Courts is left to the Patentee, whereas the idea that the State should itself police the grant it makes, is fundamental to the Warrant system.

In Chapter IV the advantages of protecting innovation directly, using either of the models described, or by combining features of both, are discussed. The first of these is that really strong monopolies could be granted, which would be socially acceptable because their length would be matched to the risks undertaken. Historically low (and falling) financial returns to investment in innovation would thus be greatly improved. Employment would be increased both by the stimulus to investment and by a shift from labour-saving to output-increasing innovations. Industry structure by firm size would be changed in favour of smaller firms, and the power of multi-national firms would be reduced. Above all, as explained at length, direct protection of innovation offers a dynamic means of meeting economic pressure from Japan. It also has advantages for Regional and Science Policies, and for the undeveloped world. Finally, the Report calls attention to the fact that recovery from Depression has always been associated with some quantum change in the law which regulates how business is done: Direct protection of innovation is by far the strongest candidate for similar legislation in our present difficulties.

CHAPTER I

THE UNEXPLOITED POTENTIAL OF PATENTS

The concept of monopoly is far older than that of patents, and must indeed be equated in age with the earliest use of political power. By giving him (or her) sole control of some valuable resource, any sovereign could ensure that a favourite could prosper. In medieval times, monopoly came to be associated with arbitrary use of power in this way, and to be contrasted with a Franchise. The difference between the two was that a Franchise conferred rights (which could be sole rights, as in the case of a monopoly) and these carried corresponding duties, but no similar duties were imposed on the holder of a Monopoly.

THE PRINCIPLE OF PATENTING

It is a little strange, therefore, that the principle of Patenting, in which exclusive rights are granted in exchange for activity judged to be socially beneficial, should have been associated with the word monopoly rather than with franchise. No Patent in the sense in which we use the word today, has ever been granted as a free gift of benefit to the Patentee, without conditions attached. A Patentee has always had to perform something *in exchange* for his sole rights. He may have had to share a discovery of something new and useful, called his invention, with the public. This would mean that he had in fact performed his side of the bargain earlier than (or at least simultaneous with) any possible receipt of reward in the form of the rights conferred on him reciprocally. Or he has to innovate an idea, that is, to take something which exists only in his own or someone else's head, and turn it into concrete reality. This will usually take place *after* the grant of a Patent. Presumably the association of Patents with Monopoly rather than with Franchise resulted from the desire to stress the fact that the rights conferred were *sole* rights. This would be the case particularly where the Patentee's contribution was innovation, rather than invention, since such rights are economically necessary for innovation to be carried through, and all the early Patent grants were of this latter type. Some of the earliest records relate to Venice, and it is clear from these that Patents granted under the famous Ordinance of 1474, gave monopoly rights to those who would bring new industries to the Republic. In the case of Britain, the Statute of Monopolies of 1624, which was a sweeping away of a mass of accumulated privileges, explicitly preserved those monopolies which had for their object the generation of "new manufacture within this realm".

All the early grants of monopolies in exchange for doing something new, were grants of Patents for *innovation*, not for invention. What the Patentee gave in exchange for his sole rights was the introduction and full working of a manufacture which was new to the country. Expenditure of personal effort and capital by the Patentee was an essential condition of his grant, and it was understood that the public would be instructed in the new technology by his personal supervision of the business, which would be a tangible example, capable of being followed.

CHANGED SUBJECT MATTER

The way in which this practice of exchanging monopoly for novelty developed historically, however, led to Patents for invention instead of innovation. Wherever Patents are granted today, they relate only to information, not to that information embodied, working, and tangible. As the means of instructing the public about the new thing, the artifact actually functioning in its environment, has been replaced by a description on paper, the Patent Specification.

Nevertheless, in spite of the fact that the subject-matter of Patent protection is now the idea and no longer the process whereby the idea is realized in practice (invention, and not innovation) innovation remains the reason why Patents exist, and their only possible economic justification. Take, for example, Article 1.8 of the United States Constitution:–

"The Congress shall have power...to promote the progress of science and useful arts, by securing for limited times to authors and inventors the exclusive right to their respective writings and discoveries".

The "exclusive right", of course, is the Patent grant, but this can only promote science and useful arts if it helps to get the "discovery" actually exploited in practice. Moreover, unless there are returns on investment for innovators there can be no rewards for inventors. By some means or other, the "exclusive right" has to be extended to the innovator's activity instead of being limited to that of the inventor, if it is not to be futile. Consequently, modern Patents still have the same ultimate object as early Patents, namely the protection of innovation. There is the important difference, however, that they protect innovation indirectly instead of directly. Invention is what is now protected directly, and it stands between the monopoly rights and the innovation. It is the inventor who is protected directly, and in the majority of cases he will transfer his Patent rights to an innovator who has the resources to turn his ideas into concrete reality and make money out of them. In fact, the ways in which this is usually done, reflect how the Patent system has changed. In Britain, a firm can apply for and be granted a Patent. Through the British system, therefore, there can still be perceived, however dimly, the origins of Patents in terms of protecting innovation directly. In the U.S., in contrast, a firm has no existence before the Patent Office, even though it is firms that are the active agents in innovation; only individuals can be granted Patents. The Patent Office ignores the reality that Patent rights are almost always "assigned" immediately to the innovating firm, to give indirect protection to an innovation.

PROTECTION AT ONE REMOVE

Since any protection given by a Patent to innovation is now at one remove, how much protection innovation receives therefore depends upon how close is the identity between the idea and its possible realization, in commercial terms. If the idea is capable of no more than a unique embodiment, indirect protection is as good as direct protection; but if it can be embodied in several ways, all of which conform to some aspect of commercial reality, the link between invention and innovation becomes correspondingly tenuous, and indirect protection of innovation moves towards being worthless.

Not the least of the paradoxes of the Patent system is that it moved away from direct, and towards indirect, protection of innovation during the early stages of the Industrial Revolution. It was precisely during this period that the evolution of technology and business was towards multiplying the opportunities for any idea to be embodied in different ways, all of which are commercially equivalent, even if not technically so. In so doing, the system made itself progressively less effective in giving protection to innovation, to the extent that it has now become almost completely irrelevant to it in many areas. Yet the principle of Patenting remains in being, and it will be argued here, that this principle can cover much more than the *existing* Patent system. Specifically, there are large and immediate benefits to be obtained from reviving the direct protection of innovation that was characteristic of Patents in their earlier form.

REASONS FOR CHANGE

Why did the Patent system develop as it did, especially if this led it towards exploiting less rather than more of the potential of the principle of Patenting, which is exchange of monopoly rights for some socially beneficial activity? The answer appears to be that Patents were never the only means which offered the protection which innovation needs. If they had been, they could scarcely have developed in any other way than that of protecting innovation directly. Historical changes, however, especially in the legal area, developed these alternative protection means to such a remarkable extent that Patents became subordinate, even secondary to them. The result has been that, today, Patents are largely a reinforcement of the protection given to innovation by these other means. The unexploited potential of the principle of Patenting therefore consists primarily in bringing the monopoly/novelty exchange back to being an independent means of protecting innovation. Even before attempting to explain the historical evolution of Patents, therefore, it is necessary to outline the other means which can confer the protection which innovation has to have, and the ways in which they interact. Patents cannot be studied on their own; they are only intelligible within a general context of market power.

THE NATURE OF MARKET POWER

The essence of a market is that anyone can come to it either with something to sell or with money to buy. To "make a market", as brokers do with stocks or commodities, is precisely to set up arrangements so that this can happen with complete freedom. Market forces are the pressures that are generated by this freedom of entry. They always work to push prices downwards. To have power over a market, then, can mean nothing else than to be able to control its essential characteristic, which is freedom of entry. The effect of that control can only be to weaken the downwards pressure on price. Paradoxically, therefore, market power is not power to *make* a market. It is power to *"unmake"* it. It is the power to escape from the constraints, especially in terms of price, which the market seeks to impose. It is always about erecting barriers to entry to a market. It acts by keeping others out. (1)

INNOVATION AND MARKET POWER

Market power is indispensable for innovation, because others simply have to be kept out if innovation is to be possible. Investment in innovation must always be at above-average risk. Consequently, it can only be undertaken rationally if there is prospect of an above-average reward. In a market where complete freedom of entry pushes price down as far as it can go, innovation is simply impossible.(2) If there is to be innovation, there must be means of interfering with the market, of erecting barriers to entry, of keeping would-be entrants out, so that price can be kept up. Market power does not enable a firm, in the technical sense, to do anything different from what it could do if it had no market power. But having market power enables it to do it more or less alone. It enables it to devote some of its resources to innovation. This demands that the firm goes into a situation as a lone investor. Market power holds out the prospect of emerging from that situation as a correspondingly lone seller, for which the technical term is a monopolist. This will give above-average returns, and this is how market power makes investment in innovation possible.

There is a second reason why market power is essential to protect investment in innovation. This is because some part of any such investment cannot fail to be directed towards the production of new information. If that information is not protected, it will become available to competitors who have risked nothing in producing it, and who will pay nothing for the use of it. Clearly, investment to produce new information without the means of protecting it, can never be a rational business exercise.

MARKET POWER AND POSITIVE LAW

There are many ways of keeping others out of the market, but all of these can be classified under one of the three headings: Capability, Persuasive or Specific. Every actual firm possesses and uses all three types, although their relative proportions differ from firm to firm. And each type of market power is rooted in some specific piece of positive law. (3)

CAPABILITY MARKET POWER

This exists to the extent that a firm has productive capacity that is specialized, or of a particular scale, or which involves an experienced workforce. Its products are therefore of a kind, and sell at a price, that can only be matched by a firm which has much the same resources. All other firms are effectively barred from entry to its market. Capability market power is always the result of investment in productive assets, and it is rooted in the legislation which enables such investment to be made on a large scale. This is in the Acts providing for Companies whose shareholders have limited liability, which were passed in all European countries in the years 1855–1882.

PERSUASIVE MARKET POWER

With the growth of discretionary income, products came increasingly to have two identifiable parts, one physical, the other psychological. As technology in any field approaches its practical limits, the physical ingredient in products becomes standardized, and the need to differentiate them psychologically becomes correspondingly important. Hence the growth of salesmanship, advertising and the techniques of marketing. All of these offer opportunities of keeping others from entering the market. The advertising which is associated with a product class becomes part of the public's definition of that type of product. Firms that can supply the physical product but not a comparable level of advertising, are not able to supply products according to this definition, and are consequently not "in" that market. The barrier to entry is the sheer size of the advertising expenditure. For the firms that are inside this type of barrier, it offers all the advantages of a cartel, with none of the problems. Just as barriers to entry of the "Capability" type are erected by investment in productive resources, at least equally strong barriers can be erected by investment in marketing, that is, in the apparatus of Persuasion. The legal root of this type of market power can be found in the various Trade Mark Acts passed in the main countries of Europe between 1857 and 1883. Without these Acts, marketing as we know it today, and especially advertising, would never have been possible.(4)

However, it is important not to limit Persuasive market power to mass market products. It also exists even in capital goods. Since decisions to buy these are influenced by individuals with career paths to follow in large undertakings, a firm that can persuade these individuals that their careers are safer if they buy its product rather than a competitor's, will succeed even though the competitive product may be technically better. In this case, persuasive market power is obtained by investing in technical salesmen to deliver this "reassurance". The firms that cannot offer it are effectively "out of the market".

SPECIFIC MARKET POWER

In this category can be grouped all those means of restricting freedom of entry to a market which depend upon some explicit and particular use of the power of the State. Licenses to prospect for and exploit minerals, planning (building) permissions, tariffs, quotas, and, of course, Patents, are all of this type. In most cases, it is easy to identify the legal root; increasingly, however, it may be necessary to trace some aspect of rule by Ministerial Order or Decree back to the empowering legislation. But the legislation will always be there.

INTERNATIONAL ARRANGEMENTS

Although they depend upon national legislation, many types of market power are world-wide in their application. Patents (Specific market power) and Trade Marks (Persuasive market power) have been internationalized since the 1883 Paris Convention. The General Agreement on Tariffs and Trade (G.A.T.T.) of 1947 does the same, although much less effectively, for Capability market power. The International Copyright Union and the Berne Convention cover the literary and visual content of advertising, and thus reinforce the protection of Persuasive market power internationally. They are now also being used to reinforce Capability market power by the protection some Courts have ruled they give to three-dimensional objects through the drawings from which these are made.

The general argument may now therefore be re-stated in the more precise form that the evolution of the Patent system was strongly conditioned by the emergence of Capability and Persuasive market power as means for protecting investment in innovation. This resulted in Patents coming to fulfil no more than a supporting role to these two types, especially that of Capability. Innovation in Chemicals is a partial exception to this general rule, for reasons which will be explained fully below.

EARLY PATENT HISTORY

During the period when Patents protected innovation directly, manufacturing was characteristically small-scale and local. Technology was based upon particular skills rather than generalised concepts, and its transfer depended upon individuals, too. Chemistry was of little importance, and the great advances in mechanical drawing which were to be essential to widespread diffusion of standardized engineering knowledge, did not come until the nineteenth century. Just as the separation of management and ownership was not to be a reality until far into the future, so inventors and innovators had not yet become separated. They tended to be the same people. "New manufacture" meant what it said, and the definitive case was that of Bircot, where Justice Coke ruled that a valid Patent could not be obtained for "a mere addition" to an existing manufacture. (5)

However, these inventor/innovators went on making improvements, and thinking up ideas which, even though they were new, did not lend themselves to exploitation in a separate establishment, as a "new manufacture within this realm". New and valuable technology, experience now showed, could be obtained from indigenous sources, and not simply by importing it from abroad. Since the Capability market power available to any of the innovators who were also probably the inventors, was limited, they naturally tried to obtain Patents as a means of protection.

Whilst this was going on, an administrative change was made in Britain, removing concern with Patents from the Privy Council to the Courts, which meant involving the Law Officers of the Crown. These Officers were then faced with difficulties arising from the new kinds of things and activities for which Patent protection was being claimed. For example, improvements to existing machines (incremental innovation) accelerated throughout the eighteenth century. The nature of these difficulties can be grasped by looking at the Patent grant from each side in turn. From that of the Applicant, the object would be to get a Patent for the widest area of activity possible; from that of the Law Officer the problem would be to have any idea at all, firstly as to the scope of the monopoly that was being applied for, and secondly, whether there was any justification for granting it.

COMPULSORY "ENROLEMENT"

A solution gradually emerged, through filing or "enrolement" in Chancery, of descriptions of what was supposed to be new. Since all questions of novelty, as well as other aspects of the Patent, were left to the Courts to decide, Patentees began this practice as a means of arming themselves with evidence against possible future litigation. In doing so, they had to be careful, in drafting their description for the purpose of strengthening their hand in the event of a Court action, not to divulge so much information as to make a free gift of the "new manufacture" to another who might pirate it. Although such descriptions as yet formed no part of the procedure whereby a Patent was obtained, the Law Officers came to see in them a means of solving their own problems of definition. From 1711 onwards, therefore, whenever they had doubts, or whenever an applicant wanted a monopoly, but was reluctant to disclose enough information for proper definition of its limits, they made it a condition of a valid grant that the description had to be "enrolled" within a prescribed period. This period varied between one and six months. Such a condition is found written into a Patent in 1723, and by 1734 it was standard practice. It has been suggested that the Law Officers transferred this idea to Patents from Copyright, since monopoly rights in books had been granted from 1709, on condition of deposit of a copy of the book.

"TEACHING", NOT WORKING, THE NOVELTY

Once the description had become an essential element in the grant of a Patent, it is easy to see how this, rather than the actual working of what was in the description, could come to be thought of as the "consideration" for the Patent grant. That is, what the Patentee gave in exchange for his monopoly was no longer the "new manufacture" as such, actually operating and instructing others in a new technology by tangible example, but the abstract "teaching" of its description as filed with the Patent application. The definitive legal judgement (in 1776) requires a Patentee to

"disclose his secret, and specify his invention in such a way that others of the same trade may be taught to do the thing for which the Patent is granted, by following the directions of the specifications without any new invention or adding of their own". (6)

From at least that date, therefore, it may be said that a Patent was granted for "teaching" a novelty, not for performing it.

Once the emphasis had shifted to the disclosure by Specification, away from actual working of the "new manufacture", Patent protection had also shifted from direct to indirect protection of innovation. From then onwards, whatever protection an innovation would receive from a Patent, would depend, firstly upon how much protection the Patent gave to its associated invention, and secondly, on how close was the link between that protection and all the commercial possibilities of innovating the "teaching".

U.S. EXAMPLE

The United States system followed a somewhat similar evolution to the British one. The first Patent Act there, in 1790, provided for examination of all applications, but this proved to be impracticable because of numbers. Examination was then only required in certain cases until 1836, when, with the establishment of the Patent Office, it was extended to all applications. When the Office started to publish the Specifications of issued Patents, another element was added which had formed no part of the original shape of Patent grants. Whether or not the original motivation of its officials in publishing them was simply to make their own searching in the examination process easier, it is not difficult to see how it confirmed the idea that giving the "teaching" of the novelty to the public so as to be published in this way, was the essential "consideration" offered by the Applicant in exchange for his Patent monopoly.

There were other ways in which the example set by the United States practice played an especially important part in ensuring that all the Patent systems which were set up later would be for the protection of invention, not innovation. Through and behind the U.S. system, it is possible to see the thinking of the Enlightenment, with its emphasis on individual creativity and the Rights of Man. Indeed, the French Patent law of 1791 is explicit on this particular point, and it is reflected in the United States case in the award of Patents only to individuals, not to firms or institutions. Perhaps even more revealing is the fact that a U.S. Patent has never had to be exploited in order to remain in force. It exists to confer rights on an inventor, and whether or not it then contributes to "new manufacture" is the inventor's business, not the Patent Office's. In contrast, the British system contains vestiges of its original function of protecting innovation directly. Firms can be granted Patents, and grants from 1575 onwards explicitly laid it down that the Patent would lapse if not exploited. This was eventually watered down to the modern provision for "compulsory licensing" if the Patent is not being worked. Although many countries make "working" a condition of keeping a Patent in force, this can invariably be complied with by a gesture that is no more than symbolic. No country insists upon the level of investment and activity that could be described as genuine innovation in any sense, and U.S. example must have been a strong influence in this direction.

OPPOSITION TO PATENTS

This then, was the situation of Patents at the outset of the Industrial Revolution. By a series of small changes, mainly brought about for reasons of administrative convenience, they had been turned into something quite different from what they had originally been. Steam power accelerated technological innovation enormously, and transformed the size and degree of specialisation of industrial units. The market power of Capability increased correspondingly, and those who possessed it were impatient of Patents, which they saw as interfering with their freedom of action. Their attitudes were typically expressed by the great engineer, I.K. Brunel, when he said that "he could hardly introduce the slightest improvement in his own machinery without being stopped by a Patent". (7)

But Brunel incarnated the forces which were expressed in Capability market power. Behind everything he did in connection with railways was specific legislation enabling huge amounts of capital to be assembled, which foreshadowed the similar but even wider effects of the later Companies' Acts. There was also the beginning of the separation of ownership from management. Partly because of this legislation, when he turned his attention to steamships, his innovations were also protected by Capability, and again he had no need of Patents.

In fact, those who opposed the Patent system and sought its abolition, over-stated their case. Study of the records shows that apart from the famous Watt steam-engine Patent, which is the exception that proves the rule, the Patent system contributed little to the huge outpouring of technological innovation of the second half of the eighteenth century and the first half of the nineteenth.

Further, those with Capability market power scarcely had need to worry about being inhibited in their operations in any serious way by Patents. A Select Committee of the House of Commons in 1829 heard evidence to the effect that half of all Patents in infringement actions were judged to be invalid. From the point of view of the evolution of the Patent system, what is important about this era is that it consolidated the earlier trend for Patents to be granted for invention instead of innovation. The existence of a class of people described as "inventors", who were not simultaneously innovators, became evident. The Patent system survived the period only to the extent that it was related to this class, since virtually all real innovations were undertaken and carried through under the protection of the market power of Capability. Patents really contributed little to the first thrust of the industrial revolution.

PATENTS SECONDARY

The next stage in the evolution of the Patent system may be described as its "domestication" in the interest of Capability. In fact, the end of the opposition to Patents can be seen in reforms of the Patent system which led to just this capitulation. The innovations of the first phase of the Industrial Revolution, in addition to being typically protected by Capability market power, were also characterized by being largely achieved through trial and error, empirically, ahead of theory. Watt may have attended Black's lectures on latent heat, but it was not theoretical knowledge that harnessed heat energy. "Thermodynamics owed far more to steam power, than steam power ever owed to thermodynamics". But the new industries towards the end of the nineteenth century did depend upon earlier scientific discoveries. Patents were undoubtedly useful for industries of this newer type, but by now, these were Patents in which the emphasis was clearly on direct protection for invention rather than innovation. Their greatest value was therefore to the continuing firm, not to a new firm, established to exploit a Patent monopoly. Other types of market power were available for the protection of risky investment in turning ideas into reality, and Patents were confirmed as being supportive of these, rather than a means of challenging them. By the time these new industries were founded, other countries besides Britain and the United States had established Patent systems, and these used Patents in a similar way.

Even though the French Act of 1844, for example, made "performing" the innovation in France an essential condition for the continuation of a Patent in force, the growth of capability market power led there, too, to "teaching" the new thing becoming more important than "doing" it, as far as Patent protection was concerned. By the time of the signature of the Paris Convention in 1883, the dominance of "teaching" as the factor exchanged for monopoly, was complete. The express prohibition in this Convention, of revocation of a Patent on the ground that the Patentee was importing goods manufactured in another Convention country, is convincing evidence of the change. The Convention came into being, of course, at what may be thought of as the high-water mark of the free trade movement. After the Depression of the late 1880s and 1890s, the changed mood favoured a reversion to "performing", but moves in this direction were frustrated, as will be discussed below.

CHEMICAL PATENTS

This is perhaps most easily observed in Germany, because of that country's lead in the new chemical industries. These relied very much on University research, and it is in the firms that comprised them that purposive R & D as an integral function of the firm, appears for the first time. Laboratories organized in this way, of course, are the workshops of employed inventors – by 1890, there were more researchers working in Munich alone, than in the whole of Great Britain. Moreover, although Patents were granted for invention, not for innovation, they gave useful protection for the innovation of Chemical inventions, because the link between a technical teaching and its commercial embodiment is strong in their case. Chemical inventions and Chemical firms were to have so much influence on the development of the Patent system, that this point needs to be examined further.

Once "teaching" becomes central to the Patent grant, then the value of the protection given depends upon how close the correspondence is between the teaching and what is actually bought and sold. For chemicals, this can be very close indeed, for several reasons. From the point of view of the completeness of disclosure required, a chemical formula is both comprehensive and exact in a way that no description of a mechanical or electrical invention, even with drawings, can be. The slightest change in any of the elements of the formula – as witness the effect of the presence or absence of a catalyst – may change the performance of the product radically. There is consequently little scope for capturing the commercial results of a chemical invention, by "inventing around" the Patent. Nor is there any opportunity for incremental changes, which are so important in the other areas where Patents are used. The only difference between the first product in the laboratory (which is perfectly described in the Patent "teaching" through its formula) and what may eventually be sold all over the world, is quantity. Even though this may be measured eventually in tons, every single unit of it is identical with the original discovery, and cannot be otherwise.

INVENTION / INNOVATION IDENTITY

Even though the Patent system which the Germans adopted, therefore, was one which protected invention directly, this gave valuable protection, albeit indirect, to their innovations, since the link between "teaching" on the one hand, and "commerical product" on the other, was (and is) a strong one in the Chemical industries which were so important to them. Patents were a powerful means of reinforcing the Capability market power of firms which could employ "inventors" and support their Research and Development.

Because of the near-perfect identity between what is invented and what is innovated in their case, it would make little difference to Chemical and Pharmaceutical firms from the strict point of view of protecting any individual discovery, whether the Patent system protected innovation indirectly or directly. However, since direct protection of innovation holds out the possibility of underwriting new rivals for firms with Capability market power, whereas protection of invention (indirect protection of innovation) reinforces existing Capability, the weight of the influence of Chemical and pharmaceutical industries is naturally thrown behind invention protection.

ELECTRICAL INDUSTRIES

The next group of industries to be developed substantially on the basis of scientific advances, were the electrical ones, which applied the theories of, amongst others, Faraday, Kelvin and Clerk Maxwell. Here again, the connection between employed inventors, facilities for purposive R&D and Patents for inventions as a reinforcement of Capability market power, is evident. All three elements can be illustrated perfectly from Edison and his firm. They were at the core of the new electrical industries, and his Menlo Park establishment was the first applied research laboratory in the world, outside the chemical field. Edison himself obtained over 1,000 Patents, and indeed, no firm could have been more Patent-conscious than his. The fortunes made by those who backed him even led financiers to a short-lived flirtation with Patents as a way of making money. During this period, for example, Elmer Sperry was able to set up several companies successively to exploit his Patents in different fields, and to obtain finance for them without difficulty.

Nothing can be more revealing, therefore, of the relationship between the different kinds of market power, than the Edison firm's explanation that it had allowed infringement of its electricity Patents to take place in 1882 because

"the business ascendancy we have acquired is of itself sufficient to give us a practical monopoly".(8)

In other words, "with Capability and Persuasive market power (for what else is "business ascendancy"?) who needs Patents!"

It was this approach that set the pattern of the future for the Patent system, in which it would have a very secondary role in financing innovation, compared with that of "business ascendancy". In non-chemical fields, it is extremely rare for a Patent for invention ever to be more than a secondary element in innovation protection. The real protection arises from Capability, often reinforced by Persuasive market power as well.

CROSS-LICENSING

Nevertheless, Patents do play an important role in the relationship between large firms, especially in facilitating cross-licensing between them. If one considers how this type of licensing might be carried on if no Patent system existed, it is clear that it would still require exact disclosure of the invention to the other party, together with a precise delimitation of what it is that is being licensed. This, of course, corresponds exactly to the Specification and Claims of a modern Patent grant, so that the Patent system offers convenient, ordered and classified documentation arrangements to large firms which would otherwise have to develop something very similar themselves for their licensing activity. To a very large extent, Patents for invention have become a second currency, cashable only by the largest firms, and used only in their transactions relating to innovation, with other firms of a similar type. If one firm has an invention which has potential outside what it regards as its own specialised territory, technically or geographically, then, of all the possible licensees, the most attractive must be one which can offer in exchange, not just royalties, but a reciprocal licence as well. Service and the transfer of "know-how" are essential supports to any licensing arrangement, and arguments as to whether poor performance is due to A's bad design or B's poor operation can be very costly to both parties. When the flow of information as well as of money is in both directions, as it is with cross-licensing, things tend to go much more smoothly, thus reinforcing the market power of the larger firms which can be members of the favoured group, as against that of the smaller firms outside. Once again, the fact that the real market power is Capability, with or without Persuasion as well, and not Patents, results in pressure for, or at the very least acceptance of, protection of innovation that is indirect rather than direct.

"DOMESTICATED" PATENTS

It is only because firms have other means besides Patents for protecting their innovations directly, that they can favour a system of Patents for invention rather than innovation. This is what is meant by "domestication" of Patents in the interest of the possessors of other forms of market power. A really effective Patent system, giving protection *directly* to innovation, would not be to the taste of such firms, since it would face them from time to time with unexpected competition from a quite new firm, founded on the basis of a Patent monopoly. With a Patent system protecting innovation directly, indeed, it would be reasonable to expect that growth would take place to a significant extent through the establishment of new firms. When Capability market power underwrites the bulk of innovation, the tendency instead is for growth to be reflected in the increasing size of existing firms. Consequently, just as the British engineers attacked the Patent system, so it is reasonable to suppose that in every country where the laws of capital deployment gave a major boost to Capability market power, the firms which were founded and which grew as a result, were antagonistic to any potential arrangements which could underwrite challenges to them. The influence of such firms would therefore have been used to cause development of Patents to move away from direct protection of innovation. It would certainly be to their advantage for the Patent system to be relegated to the protection of invention, leaving the direct protection of innovation to the other types of market power which they possessed, and which enabled them to gain most advantage from their size.

PATENT OFFICES

By the time the Science-based industries were gathering momentum, then, all countries with Patent systems had set up Patent Offices, following the United States precedent of 1836. Patent Offices occupy a peculiar and indeed unique position in the modern State. Their function is quasi-judicial to the extent that there is a sense in which they function as a sort of lower Court. Even if they form part of a Department of Trade or Industry or Commerce for administrative purposes, they nevertheless have complete autonomy within the terms of the Statute which sets them up and within which they operate. That Statute invariably gives them extremely wide powers to make Rules whose content can effectively decide where the emphasis of the Patent system is to fall. Above all, they have a monopoly in issuing Patents and this means that competition can never be the stimulus to change which it so effectively is in other areas of economic life. The only measure of their own performance, therefore, that can impinge upon Patent Office officials, is the extent to which the Office is being used. They had reason to be pleased with the statistics on this point up to the first World War, and it was inevitable that officials would come to look on the system they administered in much the same way as their largest users. These, of course, were the (by now) very large firms in industries that increasingly depended upon Science, with R&D establishments and employed inventors. Consequently, the officials did not ask whether the function of Patents was to protect innovation only indirectly, nor did they question the contemporary pattern of usage of Patents, which was as a reinforcement of, rather than an alternative to other types of market power, notably Capability. Whenever there was a question of legislation, Governments naturally looked for advice to the Patent Offices. In formulating this advice, as well as in drafting their own Rules of Practice, the Offices could not fail to be influenced by suggestions from those who used their services to the greatest extent, the Patent Agents. And their suggestions in turn, could only reflect the interests of those who made up by far the greater part of the Agents' practices, the larger firms.

INTERNATIONALIZATION OF PATENTS

The next important stage in terms of both influence on Patent Offices, and of the subordination of Patents to other types of market power, was the internationalization of Patents. This came about through the International Convention for the Protection of Industrial Property (known as the Paris Convention) in 1883. The effect of this has been to link all the national Patent systems of the world into a network of reciprocal arrangements. In itself, the Convention does not prescribe a standard Patent system for the entire world. Its basic principle is simply that whatever industrial property regime a country may have, is open to all citizens of other Convention countries on the same terms as to its own. For example, until recently U.S. firms could get patents for pharmaceutical products at home but not in Italy, whereas Italian firms could not get patents for such products at home, but could do so in the U.S. However, once the Convention was in existence, and especially once it had a permanent Secretariat, it was probably inevitable that it would become a force for standardization.

The Paris Convention is the Charter of international business, and it has played a unique role in making the multi-national corporation (MNC) possible. The MNC is most intelligible as a vehicle for translating its own innovations to a world-wide scale. Its emergence was inevitable once the Convention enabled firms to obtain the advantages of their Patents and Trade Marks everywhere, and not just in their home country.

The long-term trend which began with the Paris Convention was for the biggest users of Patents also to become the firms which operate internationally. This accelerated the development of Patents in directions which suited firms of this type.

PERSUASIVE MARKET POWER

The result was for the shape of the Patent system to be moulded even more definitively into being a reinforcement of other types of market power. It was no longer just the market power of capability that dominated, because, by internationalising Trade Marks, the Paris Convention made Persuasive market power effectively into a world-wide force also; and in doing so, raised it to an altogether higher level of importance. As will be seen below, this influenced the Patent system especially through the growth of Pharmaceuticals, the more so after the second World War with the coming of antibiotics.

This parallels how at the national level, the influence of firms with Capability market power was brought to bear in such a way as to confirm Patents in their role of protecting invention rather than innovation. This is because the effect of direct protection of innovation would be to encourage the formation of new firms which could challenge the existing ones; whereas indirect protection of innovation would lead to growth taking place in established firms.

THE MYTH OF "COMPULSORY LICENSING"

Nothing illustrates the "domestication" of classical Patents better than the history of national attempts to water down the concept of a "Common Patent Area" which was fundamental to the Paris Convention in its original form. It is evident from the new wording of Article 5, adopted at the Washington revision of 1911, that the thrust of these attempts, stimulated by the spread of protectionism, was towards forcing Patentees to manufacture locally. However, this was turned aside by the device of compulsory licensing, which first appears in the text of the Hague revision of 1925. Canada, for example, insisted upon "the establishment of facilities for working the invention" locally from 1869 until 1923, but from then onwards it operated the compulsory licence procedure.

Compulsory licensing is a very effective way of defusing opposition to Patent monopolies. It was a device used by Wernher von Siemens to persuade German legislators to introduce a Patent system, and it has been held that it was in fact a principal cause of the collapse of the anti-Patent movement of the nineteenth century. Seen in the broad context of market power, it is evident that it has many advantages for those who want to use Patents as a reinforcement of their market power of Capability or Persuasion, or both.

Its most important feature is that, whilst it does break the monopoly of the original Patentee, it does not put the licensee into a monopoly position in his place – compulsory licences are rarely exclusive. Consequently, such a licence provides no basis for investment in innovation at high risk. The worst that can happen to a Patentee, therefore, is that a situation is brought into existence where Patent monopoly is effectively nullified, which leaves the field to the other types of market power which he also possesses. Not surprisingly, then, the compulsory licence procedure is actually used only to a trivial extent, in comparison with the volume of Legislation and Rules that cover it. Apart from this, a licence cannot even be requested until three years have elapsed since issue of the Patent, and there are then many stratagems available for prolonging the period until it has to be granted in practice. Thirdly, "know-how" is far more important to-day than the Patent disclosure, for making money from an invention. Compulsion does not extend to know-how, which is in any event often protected by Trade Secret Law.

The firms which operated internationally shared this objective of keeping Patents subordinated to the types of market power in which they are strong. In addition, it was in their interest that there should be the highest degree of standardization possible in the system operated by national Patent Offices. A World Patent, after all, would be valuable to a World Company, but not to a national or regional one. The influence of the international firms, like the national ones, was brought to bear on national Patent Offices in the first instance through Patent Agents, who were natural channels for conveying the wishes of those from whom they obtain the bulk of their business. But it was also mediated to the National Patent Offices through the Secretariat of the Paris Convention, in Geneva. This Secretariat also came under pressure from forces expressing national economic interests.

COMMERCIAL DIPLOMACY

Especially from the second World War onwards, international "commercial" diplomacy greatly intensified. The advanced countries were precisely those with most advanced companies, which were the big users of the international Patent system. Naturally, the commercial diplomacy of such countries was directed towards achieving the global Patent system which would be in the interest of such firms. Within the provisions of the Convention, it would have been quite possible for Patents to be granted in some countries to protect invention, and in other countries to protect innovation. The historical reality has not been such a diversity, but a strong tendency for the Patent systems of member countries to converge.

The forces on, and the influence of the Convention Secretariat, are of particular interest in this context. Those who run international organizations have to pay particular attention to the interest of their richer and larger members, since in general, these pay a disproportionate part of the organization's running costs. Moreover, (if only because they have most, and better "experts") these countries are bound to have their own nationals strongly entrenched in positions of power in the international body. The individuals chosen for these positions are more likely to be solid and competent exponents of the existing system in their country of origin, than radical reformers. It would have been highly surprising, for example, to encounter an enthusiast for granting Patents for innovation rather than invention in the Geneva Secretariat, probably at any time, certainly in the formative years of the Paris Convention's working procedures. And just as at the national level the interests of the biggest users of the Patent system were mediated to legislators through Patent Agents and Patent Offices, so the biggest users of the system internationally had their interests mediated to the smaller countries through a commerical diplomacy which "politicised" the Convention to some extent. No matter how great the objectivity and dedication of Secretariat Officials might be, the influence of the values thus imposed on them, could not fail to have its effect. The very prestige of the Convention and the expertise of these officials would ensure this, all the more because of the lack of reflection and research in the national Offices. It would take an exceptionally self-confident head of any single Patent Office to take a stand at one of the meetings for revising the Convention procedures, which was directly opposed to the large-country consensus. We need not be surprised, therefore, that this consensus spread almost to unanimity, and a substantially common Patent system emerged throughout the world.

DUTCH AND SWISS REACTIONS

Several examples of the way in which "advanced country" and "advanced firm" interests influenced the way in which the Patent system actually developed may be quoted. Few will deny the Dutch flair for economic matters, and it is significant that they actually abandoned the Patent system in 1869, only returning to it again, under pressure from Germany, in 1907. Even to the present time, their method of administering their Patents, especially the strictness of their novelty examination, attempts to minimise the monopoly advantages which can be obtained by foreign innovators in that country. The Swiss are also renowned for their ability to weigh up the balances in any bargain. It is significant that they refused to join the Paris Convention at first, and again it was German pressure that made them do so. Nor is it any coincidence that the influences on these two small countries to fall into line with the way in which Patents were evolving internationally, came from the same source. The effectiveness of Patents for invention in the Chemical industries, and the importance of these industries to late nineteenth-century Germany, have already been referred to.

PHARMACEUTICAL INFLUENCE

Much later, the growth of the modern pharmaceutical industry gave a substantial boost to usage of Patents by large businesses. For many years, the firms which consistently recorded the highest rates of return on capital were the pharmaceutical ones. This was because they were so innovative as a result of being able to deploy all three types of market power more effectively than any other type of business could. Their products can only be produced in sterile plants, involving special buildings, skills and equipment, which give them Capability market power; those who take the decisions as to what is to be bought (the doctors) are not those who have to pay for the product, since this is either the patient or the State. Doctors are consequently exceptionally vulnerable to marketing techniques, which is why pharmaceutical firms (using their Persuasive market power) spend more on advertising and sales promotion to them, than the cost of their formal medical education; and Patents for invention give valuable protection to chemical innovation (Specific market power).

Research in chemicals and pharmaceuticals relies heavily upon scale of operations in the laboratory. Modern pharmaceutical products are the result of large-scale and painstaking testing of each new formulation against thousands of possible applications, in the hope of finding one in which the requirement of the Patent law for "utility" (an invention has to be useful as well as new) is fulfilled. The way in which Patent Offices adapt the system to the big users of their services, can be illustrated from the way this new development was treated in the Patent Law. Before 1952, the U.S. Supreme Court insisted that "a spark of genius" had to be found in every Patent application in order for it to be able to sustain a valid grant. It became increasingly impossible to keep up the pretence that anything of this type was involved in the merely mechanical screening process which lay behind most pharmaceutical Patents. The 1952 Patent Act therefore included the phrase "Patentability shall not be negatived by the manner in which the invention was made".

CANADA

A detailed study of the workings of the Patent system was prepared for the Canadian Department of Consumer and Corporate Affairs in 1976. Among its conclusions were that –
– The overall incentive impact of Patents on research is "modest".
– Disclosure of new technology through Patent Specifications is very incomplete. This is illustrated by external transfers for "know-how", measured in 1972, being no less than four times the value of Patent royalties.
– Only 6% of all Canadian Patents are issued to Canadians.

- Only about 100 inventions which were apparently judged to be worth extensive Patenting abroad, emerged from the whole of Canadian-controlled industry in 1973.
- Three-quarters of all Patent royalties received from abroad, accrued to externally-controlled firms.

Consequently, the study recommended that –

- The assumed benefit to Canada of membership in the international Patent system deserved careful re-evaluation.
- Canada should give serious consideration to the possibility of abandoning the continued maintenance of a Patent system in any form.
- There should be a new Patent law in which elements about which there was substantial certainty would be changed, as compared with the existing law. A ten year period would be provided to obtain a sound factual basis for decisions on matters which were uncertain or disputed, after which further legal changes would be made.

Massive international pressure was brought to bear on the Canadian government not to act on this Report. This included threats of withdrawal of investment in Canada. The responsible Minister quickly announced that Canada would make no radical changes in its Patent system.

IRELAND AND THE PARIS CONVENTION

Ireland is a particularly interesting subject of study in this context, since it can be regarded as one of the poorest of the rich countries, or one of the richest of the poor ones. It joined the Paris Convention in 1925, and research has shown that those responsible for the decision to join, had only the faintest understanding of what was involved. The decision was made solely by officials without reference to Parliament or even to the Cabinet! Against this background, it is hardly surprising that the Convention became the main force for shaping the development of intellectual property legislation within the country. As the Minister for Industry and Commerce put it when speaking to the Senate,

"what we are trying to do in this House is to carry out our International obligations... the better way is to legislate all that the Convention seeks to achieve".(9)

BRAZILIAN INITIATIVE

In 1973, Brazil put forward a proposal to the Permanent Committee of the Paris Convention for an "Industrial Development Patent". This would be granted in any developing country to any local firm which would actually manufacture in that country the subject-matter of a Patent granted abroad.

They were met with counter-proposals from the Convention's Secretariat for a "Technical Transfer Patent". This would be granted jointly to the original owner of the Patent in the foreign country, and to an individual or firm in the developing country with whom he had made a contract of at least five years duration for exploiting the Patent. It would lapse if manufacture had not commenced within two years of grant.

The Brazilian proposal provided that developing countries should be able to grant Patents on the original Venetian and British models, i.e. for "performing" the novelty. In contrast, the Secretariat's proposals left the economic power in the hands of the original (foreign) patentee. Local manufacture need only take place if it suited him. The "technical transfer patent" therefore relied on "teaching" as the basis for monopoly grant, and did nothing to disturb the established balance of economic power.

Although Brazil could have introduced its "Industrial Development Patent" unilaterally, it did not do so. No incompatibility with membership of the Paris Convention would arise as long as such a Patent could be obtained by foreigners as well as Brazilians on equal terms. The authority of the Secretariat must have been a factor in this, especially since Brazilian respect for the Paris Convention is particularly high, as Brazil was one of the original signatories, having ratified the Convention as early as 1884.

All the parties in these discussions were no doubt concerned to do good, and the Convention Secretariat would have tried hard, in its proposals, to reconcile conflicting interests in the best possible way. Nothing can change the obvious fact, however, that the Technical Transfer Patent proposed by the Secretariat, is in line with the interests of the big international users of the Patent system. These interests would be adversely affected if developing countries adopted the Brazilian Industrial Development Patent.

ADVICE TO UNDERDEVELOPED COUNTRIES

The same is true of the Model Patent Law proposed for developing countries by the United Nations, which is no more than a copy of the "Western" system. Those who prepared it were no doubt acting in all good faith and had the interests of the poor countries genuinely at heart. But the only Patent system they knew was the one which had evolved into a reinforcement of the Capability market power of the most advanced firms of the Western world. Their Model Law therefore mirrored this and could only lock any country which adopted it into the same world-wide system, whether or not it suited their economic circumstances.(10)

Initially, indeed, the Third World Countries themselves took the view that in becoming part of this system, they were helping their own advancement. In 1964, for example, as a result of an earlier initiative on the part of the Secretariat of the Paris Convention in Geneva, an *ad hoc* Committee of such countries, with members ranging from Algeria to Venezuela, recommended to their fellows, a Model Law which was substantially that of the advanced countries. Two assumptions appear to have been made: Firstly, that protection of inventions would reduce the time needed by the poor countries to reach the same levels of technology as the rich ones; secondly, that having such a law would encourage the flow of investment from advanced to less advanced countries. In the event, both assumptions were confounded. Patents of the classical type in Third World countries were quickly shown to be altogether irrelevant to their inventive and innovative capacities, and indeed to their needs in these respects. Compared to other factors such as the level of political stability, the presence or absence of corruption, infrastructure, education and workforce motivation, the existence or not of a Patent system was totally unimportant in any decision regarding investment. The resulting dissatisfaction on the part of the countries of this type, focussed on the Paris Convention, which therefore came to be widely viewed as binding them into a system which was not to their advantage. Since these countries had numbers on their side, even if little economic power, the replacement of the original Bureaux Internationaux Réunis pour la Proprieté Intellectuelle, by the World Intellectual Property Organization, a Specialised Agency of the U.N., in 1967, gave them an opportunity to bring pressure to bear on the Convention and its Secretariat. The resulting harmful "politicisation" process, can hardly be attributed to anything other than the basic incompatibility between indirect protection of innovation and the realities faced by Third World countries.

"ABSOLUTE" NOVELTY

Patents moved yet another step further away from protecting innovation directly, with the adoption of the doctrine of "Absolute" Novelty. According to this, the subject matter of the application must be new, not only in the country where the application is being made, but in the whole world. Nothing, indeed, illustrates the difference between direct and indirect protection of innovation better than the contrast between relative and absolute novelty as a criterion for Patentability. The early grants, by definition, required only relative novelty, because they were made for the protection of innovation directly, to encourage technology, established and understood in one country, to be brought into another. Even when the emphasis transferred to protecting invention, relative novelty remained the formal criterion for a long time – in Britain, for example, until the 1977 Patent Act. In practice, however, it became progressively eroded. Protection of invention is protection of *new information*, so novelty depends upon whether or not the same information already exists. Even relative novelty is destroyed once the relevant piece of information is accessible to the public in a particular country. The worldwide diffusion of technical information, accelerating from the mid-nineteenth century onwards, progressively narrowed the gap between information available anywhere in the world and information available locally, at least in the more advanced countries. This trend was reinforced by strict Court rulings which paid no attention to what availability of information actually meant to entrepreneurs and innovators in the real world. Typical of such rulings in recent years, was the holding of a British Patent to be invalid because of discovery in a University library of an article in a Japanese technical journal, which had never been translated. The Court held that the information in the Patent disclosure, which had been thought to be new at the time of its grant, had in fact been available to the public in Britain at that date, through the article in question. But as early as 1887, a similar judgment had been given in respect of two untranslated German Patent Specifications in the London Patent Office Library.(11) With judgments such as these, combined with the fact that the range of British libraries is so wide that little information published anywhere in the world will fail to reach at least one of them quickly, the move to absolute novelty in the 1977 Act was no more than a recognition in the law of the *de facto* position.

PRESSURES ON BUREAUCRATS

What a small country like Ireland, where the availability of local technical information can only be a tiny fraction of Britain's, was doing by making the same move as early as 1964, cannot be explained in the same way. It only makes sense in terms of the sort of bureaucratic "demonstration effect" touched upon earlier. There would certainly have been no demand for it from native innovators, who could be expected to derive benefit from Patents which would be valid if information was not available to the *local* public on the date of their grant, even though it might exist somewhere else in the world. However, since only 4% of Patents are granted to native inventors, by far the more important influences on the Patent Office would have reflected the interest of the multi-national firms, and which would have been relayed through the local Patent Agents. Firms of this type, as vehicles for spreading their own innovations around the world, have a positive interest in the adoption of the criterion of absolute novelty by every member country in the Convention. Once the Specification of a Patent Application is opened to public inspection in any country, novelty is instantaneously destroyed in all countries operating by this criterion. No locally-owned Patent can be opposed to Capability and/or Persuasive market power. In any country where relative novelty applies, the inventing firm would have to make a local application within the Convention priority period, or otherwise arrange for its invention to be disclosed to the public there, otherwise an international competitor, or a local innovator, might be able to get a valid Patent by being the first to "import" the information. If a national Patent Office consults its biggest users, therefore, about what they expect from the Patent system, it is inevitable that it will hear arguments in favour of absolute novelty. This pressure will receive subtle reinforcement from any international Conference, whether connected with the Convention or not, which may be attended by officials from such an Office. At these, the most authoritative Papers, the most prestigious and experienced individuals, will be from the countries which are the homes of the multi-national firms which absolute novelty suits. The most advanced countries have absolute novelty, the trend of the future is already in that direction (do we not all agree that a World Patent is our ultimate objective?) one is flattered to be moving amongst the world's experts, no-one likes to be considered a backwoodsman, and the line of least resistance for a small-country official is to follow the lead he is given. When his country moves from relative to absolute novelty, one more step towards making the international Patent system into a means for reinforcing Capability and Persuasive market power, will have been taken. As has been seen, subordination of the role of Patents in this way, has been an important aspect of the shift from direct to indirect protection of innovation.

"INVENTIVE STEP"

Even more effective in bringing this result about, has been the requirement for an "inventive step" or examination for obviousness. One root of this is the Prussian practice of refusing a Patent if "considerable technical progress" could not be demonstrated. Where it applies, even if the subject-matter of an application can be shown to be novel, a Patent will still not be granted if

"the differences between the subject-matter sought to be patented and the prior art are such that the subject-matter as a whole would have been obvious at the time the invention was made to a person having ordinary skills in the art to which the said subject matter pertains."(12)

It is necessary to distinguish obviousness as a factor which could render a granted Patent to be held invalid, from obviousness as part of the Examination procedure which could prevent a Patent being granted in the first instance. For example, although examination for obviousness only came into Britain with the 1977 Act, for many years previously, a ground of opposition to the grant of a Patent by a Third Party, was that its subject-matter was "clearly obvious" and an application to the Courts for a declaration of nullity, or a defence in an infringement Action, could be based upon the ground that the subject matter was "obvious" (not "clearly obvious", it will be noted).(13)

As in the case of absolute novelty, the requirement that a valid Patent should involve an "inventive step", that is, be "non-obvious" to one skilled in the art, makes Patents a reinforcement of other types of market power instead of being an alternative to them. In the first place, if no such restriction had been placed on the Patenting of anything both new and useful, these other types of market power would have been correspondingly circumscribed. No doubt Brunel was overstating the case when he complained that "he could hardly introduce the slightest improvement to his own machinery without being stopped by a Patent", but he perfectly expresses the attitude of firms with Capability market power. Those with other types of market power would naturally prefer all innovation to be underwritten by these. If they cannot achieve this, and Patents have to be accepted, their fall-back position is that Patents should only be granted for radical inventions.

In this, they find natural allies in the Examiner Corps of the Patent Offices. Members of these have to be highly qualified so as to be able to cope with the "best" inventions in respect of which Patent applications are submitted. Not surprisingly, if there is no "obviousness" restriction, they regard it as a waste of their time and talents to be concerned with a mass of "inventions" from cranks and others, which they cannot take seriously. The British Patent Examiner who himself obtained over 200 utterly useless Patents (such as for rolling snowballs down a pipeline from the Arctic to the Sahara to irrigate the desert) to prove this point, represents no more than the limiting case of such a reasonable attitude.(14)

COMMERCIAL EVALUATION

This problem, of course, is the inevitable result of giving Patent protection to invention instead of innovation. If innovation is protected directly instead of indirectly, an automatic selection process involving the expertise of businessmen and investors eliminates all the projects of least value long before they could ever reach the stage of being incorporated in a serious application for protection. If Patents are granted at the stage of ideas or concepts (i.e. for information and nothing more) there is no comparable "filter". The device which Patent Offices have adopted to make up for this deficiency is the "inventive step" examination (for non-obviousness). What this amounts to in practice is that the examiner attempts to re-build the invention out of the prior art to which he has access – generally earlier Patent Specifications. The more intensive versions of this practice are known as "mosaicing". It is the source of most of the conflict between inventors and Patent Offices, and by far the greater part of appeals to the Courts from official decisions relate to this question of "inventive step". Given that there has to be some principle by which the work of Patent Offices can be restricted to "worth-while" subject matter, the resulting conflict may be regarded as inevitable, but the grounds for dissatisfaction with the "inventive step" criterion are many and reasonable. It is a matter of common experience that hindsight is the clearest of all types of vision. Inventors and innovators have particular cause to know that what looks confused and impenetrable as a problem, will be blindingly obvious once the key to the solution has been found. The great innovator, Charles Kettering, had a motto: "The answer to this difficulty will be simple – when we reach it". The task which faces an Examiner, therefore, in "rebuilding" an invention, is totally different from that which the inventor tackled. Worse, the hypothetical "individual skilled in the art" has not even any of the limitations of a human Examiner. He is assumed to know everything relevant that has ever been published anywhere, in any language. Apart from omniscience, he has, in addition, the ability to put this knowledge together in every possible useful combination. The Patent Office examiner is no more than a real-life surrogate for this imaginary paragon, but the system makes no allowance for human limitations. A granted Patent may at any time be declared invalid by a Court on the basis of information which had not been available to the Examiner at the time of its examination.

INTERNAL RESEARCH

It is remarkable that the assumptions underlying the "inventive step" have never been subjected to what seems an obvious (in the ordinary sense) practical test. To reject an application for a Patent on the ground of lack of inventive step, is effectively to say that there is nothing in the "invention" other than the information regarding its components. Or, in other words, the very juxtaposition of these components, in itself, points inevitably towards their possible combination in a useful way. If this is so, an experienced Patent Examiner, who is the nearest thing to the "individual skilled in the art" that can be hoped for in an imperfect world, ought to be able to say what the invention is, if he is presented with the prior art on which an equally experienced colleague has based a rejection. Nothing would be easier than to carry out empirical research of this type on a substantial scale, and in fact a pilot study, external to the Patent Offices, is currently being undertaken. But the proper place for this research is within the Patent Offices themselves. The fact that they have never done it is all the more surprising when the value of the possible results, one way or the other, is considered. Nothing in their procedures causes more contention and dissatisfaction than the "inventive step" aspect of Patent examinations; if the results of the research showed conclusively that Patent examiners could, in general, identify an invention from the prior art in rejected cases, it would silence their critics on the issue. On the other hand, if they could not, it would at the very least show that these critics had a point. It would, indeed, raise a serious question as to the value of the Offices' method of sifting out worth-while from valueless inventions for Patent protection.

EXTERNAL CRITICISM

This particular failure on the part of Patent Offices to examine and understand the implications of their own assumptions and procedures, is regrettably typical. Over many decades, Patent systems have been subjected to study and to criticism, and their statistics have been analysed, but almost always by outsiders. In all the literature on Patents, it is difficult to think of a single work by a serving – or even retired – Patent Office official, that would be on an essential reading list. Outstanding officials in Patent Offices, like Bennet Woodcroft in Britain, have devoted their energies more to assembling and codifying their Offices' data, than to reflection on what the Office and the System were meant to be doing.

Such a lack of interest in research and reflection on Patents is unfortunate, since if the Offices were centres of thought concerning Patents, they would not have been so vulnerable to manipulation in the interests of their largest users. It is in the nature of all bureaucracy, however, to have its own growth as a primary objective. Where the bureaucracy is a monopoly of the State, this pressure for growth cannot be disciplined by competitive forces. Those who are substantial users of the bureaucracy's services, who provide justification for its expansion, would therefore have their interests consulted, even if they never made any positive attempt to exercise influence.

Another area which seems to call for Patent Office research is whether Examiners' criteria for the inventive step, differ by type of applicant. The sanction on an Examiner's behaviour is the fact that there is appeal to the Courts from his ruling. Since this is costly, however, it needs to be taken seriously only in the case of the larger firms. An Examiner deals with applications from private inventors and from small- and medium-sized firms, therefore, knowing that his judgment will be subject to the discipline of appeal only in the very rarest cases. In dealing with the larger firms, such appeals must always be reckoned with. Examiners would be more than human if they were not more inclined to give the benefit of any doubt about "inventive step" to larger firms, and to be stricter with smaller ones. Would Examiners' rulings on "inventive step" be different if all applications reached them in anonymous form? We do not know. It may be that Patent Offices refrain from examining "inventive step" problems in detail because they are convinced of the necessity for some means of sorting the wheat from the chaff, and no better method appears to be available at present. It is even more likely that such difficulties are inescapable, once Patent protection is being granted for invention rather than innovation, that is, when innovation is being protected only indirectly.

PATENT ABDICATION

What all the foregoing amounts to is that the principle of Patenting has so far only been exploited in a single way, as a means of protecting invention. But, just as the economic meaning of invention is found in innovation, so granting Patents for invention makes sense only in so far as they give protection for turning the idea into reality. By this essential criterion, it is now widely accepted by business men, and even more by financiers, that Patents measure up poorly.

The way in which the Patent system has developed, can quite fairly be described as a progressive abdication of direct concern with underwriting innovation. Concentration on giving protection to invention as such, was the main step which reflects this, and each successive move towards perfecting this type of protection, has involved abandoning yet another aspect of the activity of "getting new things done". For example, adoption of the criterion of absolute novelty absolves all but a very few Patent Offices of any real need to do more than rubber-stamp a search and a decision already made elsewhere.

In the nature of things, serious applicants for Patents based upon this type of novelty can only be firms which are working on the very frontiers of technology. The location of such firms will be predominantly in the most advanced countries, and it can be expected that they will make their priority applications there. The provisions of the Paris Convention will then give them twelve months within which to apply in any other Convention country. The country of first application will indeed have to carry out a novelty search and rule as to whether or not the novelty criterion is satisfied. All other countries can then rely on this. Some do so by making a requirement that a list of countries in which a Patent application covering the same invention has been made, and the results of the examination procedure in each of those countries, must be submitted. Others, Australia, for example, have specific procedures which actually depend upon the prior grant of a Patent in the U.S. or Britain.

ABANDONING INCREMENTAL INNOVATION

Just as the Patent Offices of all the largest and most advanced countries have turned their backs on a line of evolution which would have made Patents relevant to *local* innovation, so all Offices that adopted the "inventive step" requirement, equally turned their backs on the task of protecting incremental innovation. It is of the very nature of this type of innovation that it evolves out of what is already there. There is a sense in which each increment in any evolutionary process is actually implicit in its predecessor, and it, in turn, points towards its own successor. This is virtually a recipe for the kind of change that would be "obvious to one skilled in the art", and which quite lacks the element of surprise which is supposed to be a characteristic of a true "inventive step". A few Patent regimes, notably the German one, have attempted to compensate for this abandonment of protection of incremental innovation by establishing a "Petty Patent", granted without an examination for a short term. The existence of such "Petty Patents", however, does no damage to the assertion that by and large the Patent system has given up any serious pretence of protecting incremental innovation. The almost aggressive resistance to giving Patent protection to computer programs, displayed by Patent Offices, and their reluctance to bring biotechnology into the scope of the system, follow the same self-denying lines. It is almost as if, by some sort of osmotic consensus, those who run Patent Offices, having found a system that has a definable constituency amongst the largest international firms, and which can be administered more or less routinely, have decided to keep out anything which cannot easily fit into the system and its procedures.

Nothing can be more evident, therefore, than the fact that if it had been left to the Patent system to protect innovation, we would have had nothing like the quantity of it that there has actually been. For the U.S., Gilfillan plotted indicators such as number of technical researchers employed, funds devoted to R&D, and production of scientific papers, over the period 1880 to 1960. He found that with remarkable consistency, the inputs increased 340 times and the outputs by 105 times. Over the same period, Patents issued to Americans increased by only 3.3 times, about the same as population.(15) However crude the measure may be, it is evident from it that Patents cannot have been responsible for protecting more than a small fraction of innovation growth. Rational investors will not get involved in the risky business of innovation without some means of keeping "free-riders" out. This is so true that recent work has demonstrated that it is even possible to predict where innovation will take place, from the mechanisms available for capturing the resulting rewards. (16)

It is not surprising, then, that the number of Patents issued to Americans by the U.S. Patent Office, should have dropped from 27 per 100,000 population in 1967 to 17 per 100,000 in 1982. (17) It is reasonable to assume that there has been a similar trend in all other countries – except Japan, where special conditions apply, as will be discussed in Chapter IV.

INEVITABLE DEVELOPMENT

These are the outstanding reasons why the failure to exploit the principle of Patenting can be regarded as inevitable. There is no need for any hypothesis of a "market power conspiracy" to explain how Patents came to protect innovation only indirectly, instead of directly. Business men respond automatically to opportunities for making money, and if such opportunities in relation to innovation do not arise primarily through the Patent system, that system will inevitably be relegated to the sidelines. The game of investment will continue to be played without it, and it will become of interest only to specialist lawyers and scholars.

Thus, the movement from direct to indirect protection of innovation is most intelligible if it is seen as reflecting the evolution of Western culture during its "democratic" phase. An aspect of this evolution has been the replacement of privileges granted by Royal prerogative, by rights granted through democratically-controlled institutions in a framework of positive law. The problem has been that the development of this law has at no time been the result of decisions made on the basis of full understanding and appreciation of the consequences. When these consequences have been different from what was intended, the inertia which cannot be eliminated from institutions (especially legal ones) acted to slow down or even prevent the changes which were needed.

But, by definition, business men are not subject to the same inertia, and business men were helped greatly in their awareness of changes that affected them by the growth of Accountancy techniques. As long as these were primitive, it was possible to conceal the defects of the Patent system as a means for protecting innovation. Once they developed, however, it became clear to business men that Patents offered far less protection than other types of market power, and they acted accordingly. Those who were deeply immersed in the juridical and institutional functions of Patents were unable to appreciate the significance of the new types of financial measurement which had become available. Consequently, they continued (and continue) to act as if Patents were of great importance for innovation, long after they have become virtually irrelevant in many areas.

THE NEED FOR CHANGE

It may very well be that every step that has taken Patents away from protecting innovation directly was dictated by the circumstances of the time. When the primitive Specification emerged, with the result that protection was given for teaching the novelty instead of for performing it, the Accountancy expertise for making appropriate Patent grants for innovations, simply did not exist. Once there is no automatic filtering by market forces of applications for protection, introduction of the "inventive step" criterion may have been the only workable way of saving Patent Offices from wasting much time and effort on worthless applications. But the fact that the system has evolved by a series of *ad hoc* modifications does not mean that its limitations have to be accepted in changed conditions. We do not have to ignore all the achievements of Accountancy, and the techniques of measuring business risks, that have come about since the emphasis of Patents was – almost accidentally – switched from innovation to invention. The principle of Patenting, in accordance with which a monopoly is given by the State, in exchange for something considered to be socially valuable, has vastly more potential than simply granting Patents for invention. Giving monopolies in direct exchange for innovation once extended the exploitation of the principle in a different way, and could do so again. There would be many advantages in establishing such a system, but there is no point in discussing the advantages of protecting innovation directly, unless there is some ground for believing that this might be feasible. In recent years, however, two attempts have been made, quite independently of each other, to construct workable arrangements for direct protection of innovation.

Both involve a fundamental philosophic shift, from "intellectual property" to the provision of rewards for actual performance. Their features will be described in detail in the next two chapters. Chapter IV will then attempt a comparison of the performance of the classical Patent system with what could be expected from a move to direct protection of innovation by either of the systems proposed, or by an amalgam of both.

NOTES

1. cf. Kingston, W.: The Political Economy of Innovation. The Hague (Martinus Nijhoff) 1984 Ch.I
2. Schumpeter, J.: Business Cycles, London (1939) 92-6
3. Kingston, W.: op. cit. Ch.II
4. ibid. Ch. III
5. Inst. 181-3 (Coke)
6. The case of R. v Arkwright (Buller, J.)
7. Batzel, V.M.: "Legal Monopoly in Liberal England". Business History 22 (1980) 189
8. Passer, H.C.: The Electrical Manufacturers 1875-1900. Cambridge, Mass. (1953) 100
9. Senate Debates Vol. 9, cols. 568, 548.
10. United Nations: The Role of Patents in Transfer of Technology to Developing Countries. New York. 1964
11. Harris v Rothwell (1887) 4 RPC 223
12. U.S. Patent Act 1952. U.S.C.103
13. Patent Act 1947
14. Fox, B.: "Is the Day of the Patent Over?" New Scientist, 10 September 1981
15. Gilfillan, S.C.: Invention and the Patent System. Washington, D.C. (1964) 8
16. Von Hippel, E.: Research Policy 11 (1982) 95-115
17. Banner, D.V.: Jnl. of the Patent Office Soc. (1983) 586

CHAPTER II

KRONZ'S "INNOVATION PATENT"

INTRODUCTION

The application of the principle of Patenting developed by Kronz, is expressly intended to give protection directly to innovation. It sets up a sort of "innovation property" which is quite different from the intellectual property of which the classical Patent system is one type. In that system, protection of an embodiment of a concept or technical teaching, can be no more than part of, or an aspect of, whatever protection applies to the concept or technical teaching itself – it is indirect. In the Kronz system, the protection is direct, since the concept or technical teaching is not protected *per se* (that is as an abstraction, apart from a concrete embodiment). The only way of achieving protection of a concept or technical teaching in the Kronz system, would be through separately protecting every possible individual embodiment of it.

The contrast between the two types of property is reflected in their objectives. Intellectual property sets out to give a reward for ideas; innovation property to give a reward for turning ideas into concrete realities, that is, for innovative action. It also protects investment, and gives a bonus that is related to risks taken. Because it refers directly to the innovative object, Kronz argues, innovation property offers better protection of the risky investment involved, and must therefore be an improved means of promoting innovation. We must indeed honour thinking and thinkers, he holds, but we must *protect* action and those who act.

Kronz's motivation for developing his practical proposals originates from a trained Patent lawyer's experience of the traditional Patent system, both as practitioner and administrator. He found that the system of direct protection for invention, giving only indirect protection for innovation, had many drawbacks. It was complicated, expensive and insecure; it was poorly understood and misused, leading it to get a bad reputation as a protection mechanism; its effect on economic competition was slight; and it was progressively being left behind by the changes of the modern economic and technical world. For all these reasons, Kronz became convinced that it was necessary to re-establish the original "pure doctrine" of Patent protection. His proposals are particularly notable for their legal precision and their procedural subtleties. These reinforce his authority as a reformer whose criticisms arise from within the system, rather than from outside it.

FEATURES

The following are the salient features of the Kronz proposals:–

1. The object of protection is not an "invention" but an innovation, that is, the invention actually reduced to practice, and commercialised.

2. The novelty criterion could not be further away from the requirement for an "inventive step" of the Classical Patent system. See feature 12 below.

3. Anything which can be embodied in marketable new things can be protected, not just technology.

4. Processes cannot be protected directly, but receive their protection through the physical components involved in them.

5. Capacity to commercialise an innovation is just as much a condition for receiving protection as is technical capacity to realize it. Where either is lacking, it can be provided through contractual arrangements with a "substitute innovator".

6. Imported innovations can qualify for protection.

7. The territorial extent of protection can be a country, a region of a country, or by agreement, a group of countries.

8. Protection is given in the same form as in the classical Patent system, by granting a monopoly to make, use and sell the innovation for a prescribed period.

9. This period would vary from case to case, depending upon the innovating firm, the market and the project.

10. Just as protection does not apply to the invention phase, which precedes the innovation phase, neither does it apply to the diffusion phase, which follows it.

11. As with classical Patents, the limits of protection are defined by "Claims".

12. Novelty would relate only to the availability of the actual commercial embodiment to the public. It is not affected by whether or not the public may have had access to any concepts or technical teaching which might point towards such an embodiment, as long as that embodiment does not exist in a fully commercial context. Thus, novelty is destroyed only by "public prior use", whilst it is established by first commercial use.

13. Third party objections and information form an important part of the examination procedure.

14. The individual Examiner of the Classical Patent system is replaced by an expert Board which would have a strong commercial bias.

15. Examination may result in modification of the term of the grant as well as in its scope.

16. There is provision for some retrospective protection for the development work which led to an innovation.

17. An "innovation quality mark" would identify protected objects.

18. The system would either replace or supplement the classical Patent system. In the latter case, some minor changes in existing Patent procedures would be required.

19. Grants are incontestable except in so far as fraud entered into the application.

20. No fees are payable to keep the Grant in force.

21. There is no obligation to continue use after the first act of commercialisation, but this can result in substantial loss of rights.

22. The Patent Office would be actively concerned with all licences granted, and these licences would also be for variable terms, reflecting the licensee's characteristics and other factors.

THE SUBJECT-MATTER OF INNOVATION PROTECTION

In the Kronz system, what can be protected is an artifact whose use is new within the geographical limits to which the Patent applies, in the form in which it is actually put on the market. The geographical limits may be smaller or (obviously, by agreement) larger than a country. None of the stages which necessarily precede the first commercial act – origination of a concept, discovery, design, models or prototypes – qualify for protection in their own right. However, they may receive a degree of retrospective protection as a result of the eventual grant of protection to the innovation in its commercial form. Kronz recommends the replacement of the phrase "technical object of innovation" with its connotation of something static, by "act of technical innovation". The more dynamic overtones of the latter would correspond to the "act of commercial innovation" which complements it, and both would stress the capacity of Innovation Protection to cope with the fluctuating realities of economic change.

Kronz's proposed text for the main Clause of a law establishing the new kind of protection is:–

"A Patent must be granted for an innovation which has not been in prior public use in the "home territory" on the day of the Provisional Grant of the Patent. An innovation shall be understood to mean the set of facts constituted by the immediately utilizable embodiment of the object of the innovation and the first commercial exploitation thereof".

FILLING GAPS LEFT BY CLASSICAL PATENTS

It is important to understand that although the Kronz system refuses protection to "inventions", it is ready to grant it to many "inventive teachings" if they are in a tangible innovated form, which could not obtain classical Patent protection. This would be because they would be judged to have too low a level of invention, or to lack an "inventive step". Such cases might involve:–

Transposition: Where a technical solution to a problem in one area of technology is also shown to be applicable in another. For example, a means of drilling holes in some synthetic materials might also find a use in dentistry.

Application: One technology may be analogous to another. For example, a way of handling glass bottles could be used also for plastic ones.

Identification: Where the supposed merit of an invention consists in discovery of a problem to be solved, rather than in the solution found to it.

Function: Formulation of an operating relationship between specified elements into a rule which can have a wider applicability.

Selection: In which the essential element consists in selecting the appropriate components out of a large number of possible ones.

Simplification, Combination, Aggregation are other activities which often find it impossible to obtain protection from the classical Patent system, on the grounds of lacking an "inventive step".

Any of these, however, when embodied in a concrete form, could find protection under the Kronz system. In this, the innovation object is anything new, but only in the form in which it actually enters into commercial activity. Moreover, protection extends beyond the individual object that is actually sold, in two ways: Firstly, copying it merely by substituting "technical equivalents" is barred. Simply changing a component, material, the scale of operation, form, proportions or arrangements for powering the innovation will not escape the protection of an Innovation Patent.

Secondly, the Patentee is allowed to list and describe variants of his innovation other than the one he has actually preferred to use in the market-place. The protection he will receive for these will not be as complete as for the version he has actually adopted. Others will be allowed to make and market them if they wish, but only by paying him a royalty. The emphasis on commericalisation in the Kronz system is evident from this approach. It forces an innovator to select, out of all possible variants of his ideas to turn into concrete reality, not the one which is technically most elegant, but the one which will best meet the market's requirements.

SIMILARITIES TO COPYRIGHT, PLANT PROTECTION, TRADE MARKS

Kronz calls attention to the way in which his proposals parallel how protection is granted for creative works by copyright, and for new plant varieties. There is a sense, too, in which Trade Marks only protect what is actually commercialised.

In the case of copyright, protection is not given to any idea for, or concept of the work which an author or artist might have had in mind. Copyright protection is attached specifically to the work itself and covers the form in which, from the moment at which, it actually comes into being. Moreover, this type of protection does not depend at all upon the presence or absence of any particular quality in the work; there is nothing, for example, in Copyright protection that could be compared, even by analogy, with the "inventive step" or "novelty" criteria of the classical Patent system. What matters is that something concrete has been produced, as the result of original effort.

Similarly, plant breeders' rights only come into being when field tests show that a genuine new variety has really been brought into existence – not when someone has an idea for such a variety, or at any stage earlier than that of complete development. Also, Trade Mark registration protection can be lost in many countries if the Mark is not "used" within comparatively few years. In other words, there must be a "commercial act" involving it. Correspondingly, there must be a "commercial act" as well as an "innovation object", if innovation according to Kronz's definition is to take place. Valid grant of an innovation patent does not require that the patentee be involved in anything other than the *initiation* of the innovation phase, once he has completed the last stage of the invention phase. He does not have to continue the innovation after the first act of commercial exploitation, although, as will be seen below, he loses some protection by failing to do so.

The "initial commercial act" would be defined by statute, and Kronz lists sales promotion, showing at exhibitions, commissioning plant with a view to production, re-casting of operations to take account of the innovation, or supply to distributors, as typical acts which would qualify. Note that an offer to sell constitutes the "commercial act", even before an actual sale is made.

Also, internal use qualifies, as long as it takes place within a commercial firm, since this is considered as having consequences in the commercial world outside. Use within a public research laboratory would therefore not qualify for protection, which could only be obtained if the laboratory brought in a commercial firm as substitute innovator, to carry out the "commercial act" which it is unable to perform itself.

Military innovations, or innovations that are excessively harmful to health, are specifically excluded from protection in the Kronz system.

APPLICATION TO PROCESSES, ETC.

The innovation object must be in the "ready to use" form. Kronz insists on this on grounds of principle, but it is also necessary to the way in which processes (whether chemical or other) receive protection in his system. It is clear that a process, considered as a guide for technical action, prescribing steps to be taken, operating procedures or values, instrument settings and the like, cannot be a tangible object, and only tangible objects can receive innovation protection. Since this is in fact granted only to the combination of tangible object and the initial act of commercialising it, processes as such cannot qualify.

To bring them into his system, Kronz uses an analogy to the "product by process" protection of the United States Patent system. In his case, this results in protection of a process through its components. A listing of both the hardware and the software involved in a process (e.g. specific settings must be supplied for equipment which can operate in more than one way) results in a description of the process in a fully "ready to use" form, in terms of all the tangible objects involved in it. A similar approach permits innovation protection to cover many areas which are excluded from patent protection at present. These include printed matter, methods of doing business, and computer programs – anything, in fact, that can be described by using a "vocabulary" of tangible objects, and which is capable of being commercialised. The objective is to provide protection for anything which can provide a nett contribution to Society. The Innovation Patent Statute, or Regulations made under it, would designate the areas of economic activity in which protection could be obtained.

It is normal Patent practice that the Specification acts as a "Dictionary" for the Claims. That is, the precise meaning of each word, as used in a Claim, is defined by the sense in which that word is used in the Specification. Kronz arrives at the "meaning" of a process by describing it as an aggregate of tangible objects, interacting together. "Settings", "readings" etc., of particular tangible elements, define the precise meaning of the words used to describe each element in this particular context, just as the Specification does in the classical Patent system.

In the case of a chemical process, for example, the apparatus used would be described, as well as the substances according to their physical state, as these are used in the apparatus. The "settings", "readings" or "timings" of all the interacting components of the apparatus would be given, as well as their mode of interacting, and the inputs and outputs of the operation would be given in terms of energy and materials. The same approach could be used in respect of instrumentation (for automatic control of an aircraft in flight, for example); of the various steps in construction of a building; of horticultural and agricultural activities; or of genetic engineering.

THE "SUBSTITUTE INNOVATOR"

There is an important corollary of the principle that protection may only be granted to a combination of innovation object and commercial act of use. This is that innovation protection can only be given to a firm, not to an individual, since it is only a firm which can effectively commercialise a product. Nor can it be any kind of firm: If it is unable to take *both* steps, of devising its innovation and of commercialising it, it is not a fit subject for the grant of innovation protection. This restriction is necessary to Kronz's approach, in which the actual innovation, realized in its correct economic context, is the fundamental object of the grant of innovation property.

At the same time, he is conscious that such a principle could eliminate from the range of protection much activity and effort which is properly described as innovative rather than inventive.

' Activity of this type can be undertaken by public research laboratories, by firms which are technically competent, but which lack the resources to take the risk of actually putting a product on the market, or even by private individuals. The device which Kronz uses to bring such agents of innovation into his system is the "substitute innovator". This is a firm which possesses whatever resources the Applicant lacks, either in terms of technology or for commercialisation. The resources of such a firm would then be made available to the project by means of a contract, the resulting combination of firms having both the technical capacity and the commerical resources to put the innovative object on the market, and so obtain innovation protection. The Patent could be granted to either of the two partners, or to the partnership.

NEW TYPE OF PATENT OFFICE

Kronz envisages that his system would be operated by a semi-public Institution. This would involve Chambers of Commerce, Trade and Industrial organizations and similar bodies which are actually concerned with business. Compared with the existing Patent Office, the change in emphasis will be reflected especially in the approach to the examination of applications. The single Examiner of the classical system will be replaced by a Committee of Experts, on which representatives of practical business will be strongly represented. Bodies such as the Chambers of Commerce will be especially important for evidence of novelty "in the market", since the requirements for a Provisional grant of an Innovation Patent in this respect, can be met automatically and immediately by a Certificate from such a Competent Authority.

VARIABLE TERM

Each applicant would have to submit separate plans for investment, production and sales, to the Patent Office. The Office would keep these plans secret, but would make use of them in assessing the "innovative capacity" of the applicant, and the likely evolution of the market. Both these factors are taken into account in settling the term of the monopoly to be granted, once novelty has been agreed. The term runs from the date of definitive grant, and there is no provision for any extension of it.

The fact that "innovative capacity" has a bearing on the term of the Patent, opens up the possibility for two firms to combine to obtain a mutually beneficial "term". This can be separate from "substitute innovator" arrangements. For example, a proposed innovation might be interesting to a large company, were it not for the fact that, because of this company's substantial innovative capacity, the term of the monopoly which it will be awarded, is considered to be too short for the profit the innovation needs to earn.

"CO-PROPERTY"

Such a firm, however, may be able to find a small firm for which the innovation would also be an interesting development. Because this firm's "innovative capacity" is low, its "term" would be correspondingly long. If both firms agree, the patent application can be filed, and the patent granted, as a "co-property", meaning that it is shared equally. The term which will then be granted is the arithmetic mean of the terms which would apply to each firm separately. Clearly, both would benefit. The large firm would have given up half the ownership of its project, but gained a longer term of monopoly; the small firm would have given up no more than a notional part of its "term", since it was not the originator of the innovation. Moreover, it is being brought into dynamic contact with the technology of the larger firm, which is likely to be more advanced than its own.

The same principle would apply to a private inventor. He would, of course, have to recruit a "substitute innovator", but once he has done this, because he himself has little or no "innovative capacity", his "term" is correspondingly long. This gives him something more to offer to the firm which is his "substitute innovator", than just his ideas. On a co-property basis, the effective term will be very much longer than this firm could obtain for the innovation on its own. Kronz proposes that the innovative capacity of a private inventor should be established by comparing his annual earnings with the cost of one production unit of the innovation object. This would give some measure of the value of his time in terms of the market to be served.

The "co-property" approach, Kronz suggests, should also apply to those who provide venture capital. A firm which is qualified to carry through an innovation in every way except for financial resources, has to find a substitute innovator. In this case, since the technical expertise is already available, such a substitute will consist only of a financial partner. Although it is not made explicit in Kronz's writings, it can be assumed that the innovative capacity of a firm which has money but no technical expertise, would be rated low by the Patent Office. The arithmetic mean between this term and that of the technically qualified firm, will thus be sufficiently high to raise the attraction of investment in innovation for both.

INNOVATION DIFFUSION IS NOT PROTECTED

Just as Kronz does not consider that the inventive step, which precedes innovation, should be protected, neither would he give protection to *diffusion* of the innovation, which succeeds it. This raises the question, when does innovation end and diffusion begin? Kronz believes that the answer is to be found in the time-scale of the *returns* which can be obtained from investment in an innovation. Clearly, the minimum term should permit the recoupment of the capital invested in the innovation. An upper limit could be imposed by ending the term for which the monopoly is granted when the diffusion stage is reached. Kronz suggests that a maximum term should be defined as the time necessary, with the production capacity of the original investment, to produce enough goods to satisfy half the total demand for the innovation object (or the objects manufactured with it). This he takes to correspond with the innovative phase; satisfying the remaining half of the market is diffusion of the innovation, and consequently not entitled to protection.

INTERRUPTION OF INNOVATION PHASE

It will be recalled that an applicant need only *initiate* the innovation phase in Kronz's system, but is not bound to continue it. In spite of this provision, Kronz accepts that application for, and acceptance of, the, monopoly grant implies the intention to continue production and sales of the innovation object throughout the innovation stage, i.e. at least up to the commencement of the diffusion stage.

If, therefore, without good reason, the patentee ceases any aspect of this continuing process of bringing his innovation to the diffusion stage, in those cases where he has been considered at the time of grant to have the required innovatory capacity on his own, he will lose his monopoly for whatever he has discontinued. In cases where, because of lack of some element of innovatory capacity, the patentee had been required to involve a substitute innovator to make up this lack, the Patent Office will first call formally on this substitute innovator to continue whatever action has lapsed, before finally withdrawing protection. If, after a period of total interruption, it is desired to revive the innovation, the patentee's term is reduced by the length of the interruption period, subject to the provision that no such deduction shall reduce a term to less than half of its original length.

FILING PROCEDURE

In the Kronz system, it is only *commercialised* innovation that counts for protection, and that protection can only commence definitively some time after the first act of commercialisation, because of the way in which the Examination stage necessarily intervenes. Consequently, there has to be some provision for *retrospective* protection for the period when innovators are particularly vulnerable, when others know of their innovation, but before their grant. Filing of information concerning the innovation as this moves towards commercialisation, is the means of providing this retrospective protection in the Kronz system.

Kronz is even prepared to consider that everything reduced to practice by the Applicant before the Provisional Grant of his Patent, and notified to the Office without a request for publication, could not be copied. This would replace the prior internal secret use of the classical Patent system by prior secret *notified* use.

At any time, a potential innovator can file in the Office whatever information he wishes, and even define the shape of the protection he hopes to obtain, in the form of Claims. All information submitted to the Office will be kept confidential to it, or published, at the Applicant's option. It is in the interest of the applicant to publish as he goes along, since this gives him "prior use" rights, if and when his Innovation Patent application is granted.

Apart from the provision of retrospective protection in this way, Kronz is of the opinion that the phase of developing and making prototypes, pilot processes and text materials, is also proper subject-matter for Copyright protection. This is because such embryonic embodiments of the final innovation objects are three-dimensional versions of plans or drawings in two dimensions, and these are certainly protected by Copyright. Moreover, they are no more than transient embodiments of such two-dimensional representations, produced solely for the purpose of learning (all innovation is a learning process). Since this precludes them from functioning technically according to the *final purpose* of the innovation object (which is to satisfy some commercial demand) Kronz holds that they are "within the realm of Copyright law". Consequently, in his system, Copyright and the Innovation Patent provide a continuum of protection, right through from idea to product on the market.

EVOLUTIONARY DISCLOSURE

In the classical Patent system, an application has to be complete on the day it is filed, and the subsequent introduction of "new matter" to the disclosure is absolutely prohibited. In the Kronz approach to innovation protection, in contrast, the "disclosure" is intended to be built up step by step, by lodging documentation in the Patent Office as each new problem encountered in developing the innovation towards commercialisation, is overcome. An application may begin with a single sentence, stating what the innovation is intended to be, or it may be an elaborate research study; whatever the first or subsequent documentation is, it is accepted for filing (and publication if the Applicant requests it) by the Patent Office, which otherwise takes no action on it.

There is an analogy here with the Document Disclosure Program of the

U.S. Patent Office. Since in the event of a dispute (called an "interference") the U.S. system gives the Patent to the first to invent rather than to the first to file, an inventor may lodge evidence relevant to invention date with the Office up to two years before his actual Patent application. Just as this will be taken into account if there is an "interference" between Patent applications in the U.S., so in the Kronz system, all documentation lodged earlier comes up for consideration at the Examination stage. It should reflect "an uninterrupted sequence of innovative action", and if it is desired that any protection eventually granted should be retrospective, it must include Claims.

The intention is that this documentation will eventually grow into a description of the innovation object which is finally launched commercially, that is so comprehensive and detailed as to allow an inter-disciplinary team of average persons "skilled in the Art" to copy the innovation in every respect as soon as its protection has expired. This description, of course, is the Specification of the innovation which will be published by the Office if a grant is made.

CLAIMS

Kronz favours a listing of all the characteristics and elements in the innovation in the Claims, according to existing national and international Standard specifications and catalogues. Use of these, he believes, would also contribute to the evolution of such publications in a useful way, as innovation objects became incorporated in them. He does not exclude the use of classical Patent-type claiming procedure, if this is "peripheral", as in the "Jepson-type" Claim in the U.S. system. A third approach which he thinks might be adopted, is marking of all the items in the technical drawings in relation to the complete specification. To illustrate the method he prefers, Kronz gives two simple examples: These follow traditional Patent practice by first enumerating the items which are not new, and then specifying those that are:–
EXAMPLES:–
RACING BICYCLE
PART I   A frame of the type X_1, wheels of the type X_2, steering of the type X_3, brakes of the type X_4, driving mechanism of the type X_5, gears of the type X_6, lighting of the type X_7,
Characterized by the following elements:
PART II   The reverse gear Y_1, the spokes' arrangement Y_2, the saddle design Y_3, the Zig-zag geometry of the steering Y_4, the pedals Y_5, all non-rotat-

ing elements made of plastics, the saddle being disposed exactly in the centre of gravity of the cycle, all items being claimed separately and in combination. (The references X and Y respectively, would refer to the drawings of the Specification).

TREE-PLANTING SYSTEM

PART I — Trees of the type X_1, instrument for hole-digging X_2, liquid manure of the type X_3, armature for stabilising planted tree X_4,

Characterized by the following elements:

PART II — Item X_3, at a temperature of Y degrees, pit in diameter smaller than that of the roots when extended, armature anchored on ground surface, tube inserted into the ground down to the roots level, man or machine provided for filling manure through tube every two hours.

(NOTE: A "process" is invariably expressed in terms of a programmed man or mechanism, because in any process there is in fact a need for either men or a device or both. Tangible things, therefore, in their "programmed" form, can enable protection to be given to processes in the Kronz system.

EXAMINATION OF APPLICATIONS

In examining an application for an Innovation Patent, therefore, many different dates may have to be considered, when evaluating the Claims in the light of the evidence advanced in opposition proceedings. The nearest parallels in the classical Patents, relate to multiple "Provisional" applications in Countries which follow the British model, or multiple applications for which priority is claimed when an application is made under the Paris Convention. In both cases, each individual Claim is assigned a particular date, which is the date on which the disclosure which justifies it, was made. Since every application under the Kronz system can begin with a minimal statement of intent, and must end with a comprehensive description of the commercialised innovation object, each of the interim stages in the build-up of documentation, can have a corresponding claim, and each such claim will have its appropriate date. Thus, the different categories of interaction between an application and the evidence submitted to prevent or limit its grant, are rarely concerned with an Innovation Patent application as a whole. Each Claim will be considered individually, and will be eliminated or modified according to the evidence which is applicable on its particular date. However, the date of the definitive claim, called the "Copy Claim", is the commercialisation date.

EXAMINATION FOR NOVELTY

When an applicant can supply proof to the Office that an initial act of commercialisation of the technical innovation object covered by his documentation, has taken place, he can apply for a provisional grant of protection. A declaration by a competent authority, such as a Chamber of Commerce, that the subject-matter of the application is indeed novel, will be enough for such a grant to be made immediately, with effect from the date of the initial commercial act. The Office will then publish the Applicant's Specification so that the provisional grant can be opposed by any interested Third Party.

Since a grant is irrevocable, once definitively made, unless it has been obtained through fraud, it is clearly very much in the interests of all competitors to submit at this stage, any information concerning prior use which they can obtain, and which could defeat the applicant's claim to novelty. As an added incentive to them, Kronz suggests that no provisional grant of protection should be valid with reference to those who are genuinely involved in its opposition. Although Third Parties are expected to produce the bulk of the necessary information regarding novelty, the Patent Office will nevertheless carry out its own independent investigation. Frivolous opposition will not be considered at all.

It is of the essence of classical Patent examination that a single document on its own can defeat a claim to novelty, if it can indicate the prior availability of a concept or teaching. In the Kronz system, such a document would carry no weight at all, since protection is not being given for a technical teaching, but for *embodiments* of teaching. A document would only carry weight in the examination for innovation protection, to the extent that it provided evidence of prior reduction to practice of the concept or teaching *together with* its actual use in public.

EXAMINATION

For the scope of the claims

In every type of protection, the boundaries of the grant have to be defined precisely, so that others may know what they are free or not free to do. Both classical and Kronz systems express these boundaries in terms of Claims. Kronz calls his particular claims either "Copy" or "Option" claims.

"Copy" claim

This single claim covers the innovation object in its precise concrete details, successively itemising its elements, features and components, in a similar manner to what is known as a "Jepson-type" claim in the United States Patent practice. As with classical Patents, the protection defined by a "Copy" claim extends to technical equivalents.

"Option" claims

These cover alternative and tested variants of the actual innovation object which has been the subject of the "first commercial act". Again, description must be in concrete detail. "Option" claims would have weaker legal force than the "Copy" claim, and Kronz envisages as one aspect of this, that a licence could not be refused, if requested. Since under his licensing procedure, as will be discussed below, the Office has to be involved in issuing a new grant to any licensee, which would carry his own individual "term", it can be presumed that it could also act as arbitrator regarding the fee to be paid for a licence under an "Option" claim.

Kronz holds that his use of "Copy" and "Option" claims is particularly suited to the "non-technical" realities of animals, plants, genes, vectors, plasmids and micro-organisms. Where there is living matter ready for commercial use, combined with novelty of use, there is an "innovation object" according to the Kronz criterion. The main (copy) claim refers to the concrete embodiment. ADN sequences are nothing else except instructions for copying this embodiment, so the act of using such sequences is one of copying, and therefore would infringe a valid main claim. Since natural multiplication is biological entities copying themselves, the main claim also covers this multiplication. Kronz's system therefore protects all the results of purposive "breeding", spontaneous mutations being covered by the "option claims". Note that the content of an Option claim *per se* cannot be cited against the novelty of another Innovation Patent application - it only becomes effective once it has been embodied.

ANTICIPATION

In the Kronz system, there are in principle four situations in which an application for an Innovation Patent could be "anticipated" by earlier acts (never, of course, by earlier "ideas", "concepts" or "teachings"). Of these, the first pair are called by Kronz "actual" anticipation, and the second pair, "notional" anticipation.

Anticipation is described as actual when the anticipatory act takes place before the filing date of a particular application, so the Applicant could know about it. Notional anticipation is when the Applicant could *not* have known about the anticipatory act at the filing date. The diagrams on the following pages are Kronz's own method of illustrating the differences.

It should be clear from the diagrams that actual anticipation might be the public use of an innovation which, for whatever reason, is not the subject itself of an Innovation Patent application. Or it might be the public commercial use of an innovation for which innovation protection is being sought, in which case the provisional grant of a Patent may also have been made before the new application. If the technical content of the anticipatory act is identical with that of the Innovation Patent application, no Patent can be granted on that application. If there is only partial overlapping or interference, then the Claims of the Innovation Patent application must be restricted accordingly.

In contrast, "notional" anticipation could be the public use of an embodiment of a concept which might not be "commercial" before the applicant's filing date; or it might even be the first commercial act in respect of another application for an Innovation Patent, where the filing date of the other application is earlier than that of the first one, and the second applicant had instructed the Patent Office to keep his documentation in confidence.

Such cases are treated in the same way as actual anticipation when there has been no interruption of the "sequence of innovative action" by the performer of the anticipatory acts. That is, identity of technical content will mean that out of two applicants for an Innovation Patent, only the first, both to apply and to commercialise, will obtain a grant. If the notional anticipation involves public use without an associated application for an Innovation Patent, identity of technical content will prevent the applicant from obtaining a grant. In both cases, partial overlapping, or interference, will require modification of the Claims of the application whose public commercial use is later than the act or acts which anticipate it.

If the "sequence of innovative action" has been interrupted, however, no prior public use can be available as an anticipatory act, and therefore no application for an Innovation Patent can be affected. In other words, it is not open to a firm which embarks upon a project but then drops it, to revive it when it sees that a competitor is following the same line, in the hope of restricting the scope of that competitor's Innovation Patent, if it achieves technical success. Throughout, the Kronz system sets out to reward successful innovative *action*.

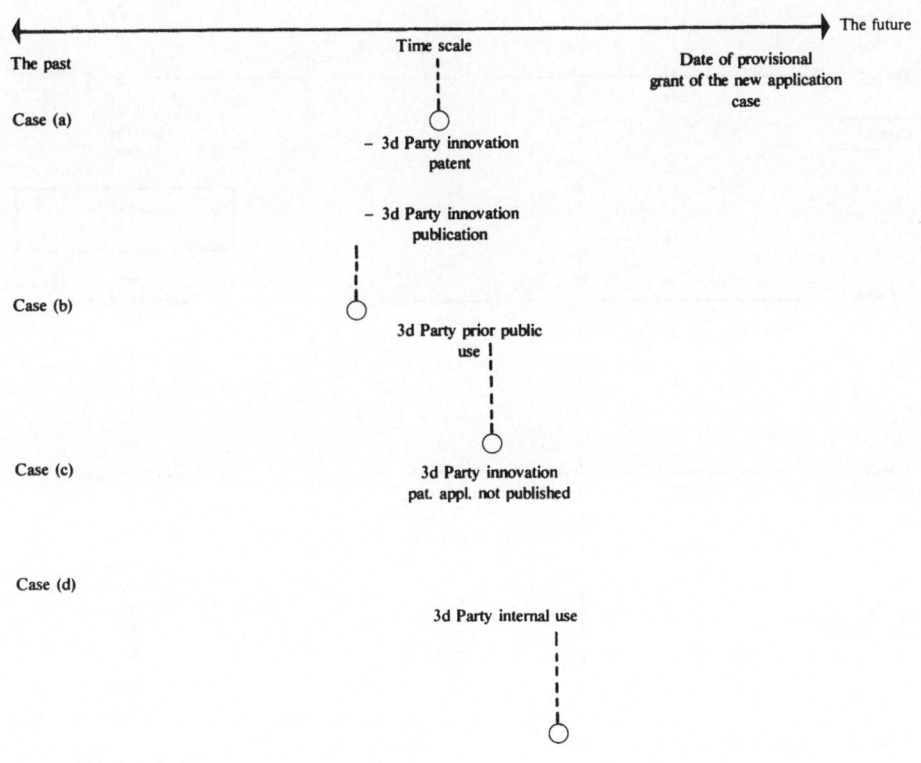

The past

The future

Time scale

Date of provisional
grant of the new application
case

Case (a)

‒ 3d Party innovation
patent

‒ 3d Party innovation
publication

Case (b)

3d Party prior public
use

Case (c)

3d Party innovation
pat. appl. not published

Case (d)

3d Party internal use

(The positions of ◯ have been chosen arbitrarily)

Graph 1: The four prior art situations globally.

52

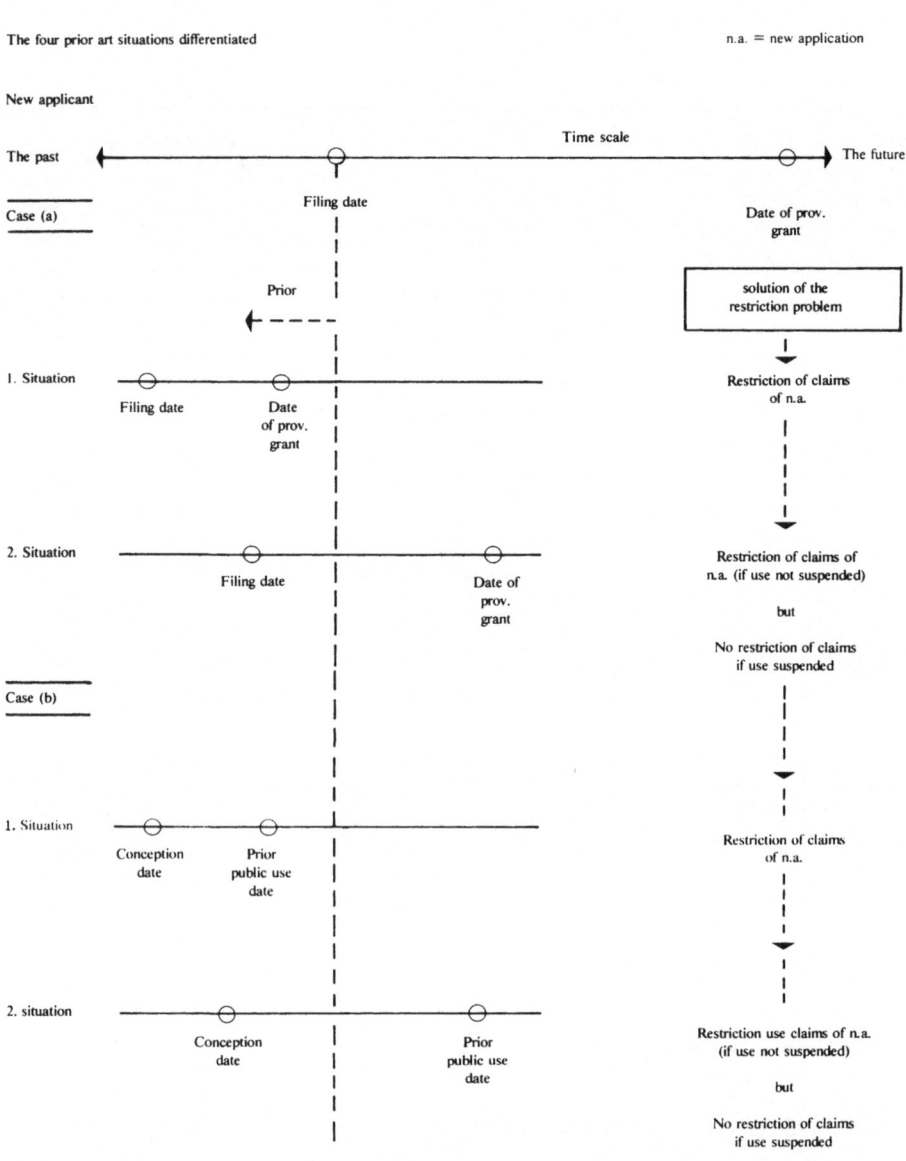

The four prior art situations differentiated

n.a. = new application

New applicant

Time scale

The past — Filing date — The future

Case (a)

Date of prov. grant

Prior

solution of the restriction problem

1. Situation

Filing date — Date of prov. grant

Restriction of claims of n.a.

2. Situation

Filing date — Date of prov. grant

Restriction of claims of n.a. (if use not suspended)

but

No restriction of claims if use suspended

Case (b)

1. Situation

Conception date — Prior public use date

Restriction of claims of n.a.

2. situation

Conception date — Prior public use date

Restriction use claims of n.a. (if use not suspended)

but

No restriction of claims if use suspended

Graph 2. The four prior art situations differentiated.

New applicant

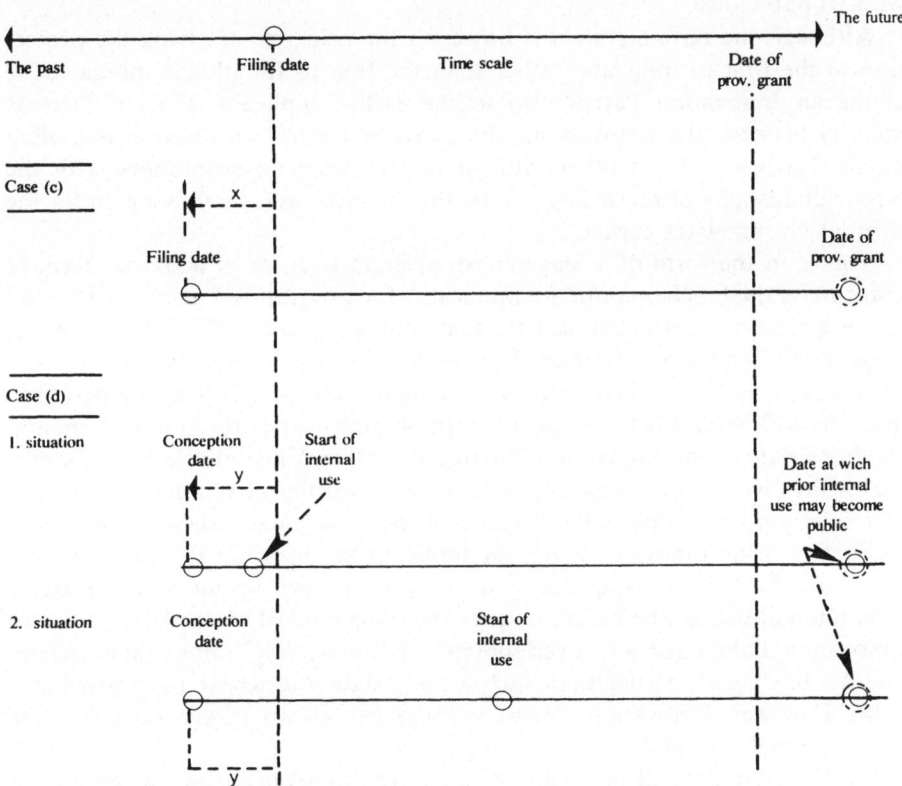

The past   Filing date   Time scale   Date of prov. grant   The future

Case (c)

Case (d)

1. situation

2. situation

Note: Conception date may be the date where first evidence of the innovation
project has been fixed.

This is also evident in the treatment given to the third category of interactions between cases involving more or less contemporary innovations. In this, because of luck, or superior R & D capacity, a firm which starts later, may reach the "public use" stage earlier than another. Where both are applicants for Innovation Patents, the period between filing and Provisional grant of one, could be completely within the corresponding period of the other. Where only one is an applicant for an Innovation Patent, the period between its filing date and Provisional grant date could be entirely within a period which starts with the "conception" of the other innovation, and ends with its public use.

Although the Kronz system is based on the principle of giving the protection to the first to innovate, rather than the first to file, this is mitigated by giving an Innovation Patent also to the earlier applicant. Even if there is identity between the innovations, the scope of neither's protection is limited by the existence of the other. Still, it is completely in accordance with the Kronz philosophy of rewarding *action*, that in such cases there is a gain for the firm which innovates earlier.

This is in the form of a longer term of grant to it, or of a shorter term of grant to its rival. Where both are applicants for Innovation Patents, the period to be added or subtracted will be that separating their filing dates. In the diagram, the time to be subtracted (or added) in case (c) is indicated by "X"; in case (d) it is "Y". When only one is an applicant for a Patent, the period in question will be that between the conception date of the other innovation, and the filing date of the Innovation Patent application. This reflects Kronz's view that a firm which innovates in public, is more worthy of protection than one which only does so internally, because of the value of so-called "show-how".

Further, if an innovator is not an applicant for an Innovation Patent, it is likely that he will be using the innovation in his own business. The start of such internal use can be before or after the filing date of an application for an Innovation Patent for an "overlapping" or "interfering" innovation. Where internal use begins earlier than such a filing date, the Kronz rules provide:–

(a) That any Innovation Patent granted carries no power *vis à vis* the internal use; and

(b) That the internal user cannot attack the Innovation Patent, no matter what its Claims contain.

Kronz foresees the possibility that an applicant would be permitted to "claim" the priority of a certified scientific discovery if this has a causal relationship to his own innovation object. Clearly, if this discovery had not been made by the applicant himself, such a "claim" would require him to pay royalties to the discoverer. The advantage to an applicant of making such a "claim" would be that it would improve his position in terms of "priority" in cases where anticipation is alleged against some part of his application.

INFRINGEMENT

Protection is granted for the combination of the innovative object with its commercialisation. It follows, therefore, that any act of infringement of the protection must contain *both* these elements. Manufacture of all the parts which, when assembled, would make an object which carries Innovation Patent protection, does not, in itself, constitute full infringement. Neither would the act of simply selling the innovation object. Each is called by Kronz, "indirect infringement". Only when both types of indirect infringement are combined, is there an act subject to penalties.

LICENSING

In classical Patent systems, the Patent Office plays no, or almost no part in licensing. In the Kronz system, the Office is so deeply involved that it actually specifies the term of a licence. The reasoning behind this is that all monopolies granted by the Office should be proportional to the innovative capability of the grantee. This is considered desirable for optimum innovation. This objective, however, would be defeated if a monopoly granted to one innovator on the basis of his innovative capacity, could be transferred by licence to another firm which, if it had been the original applicant, would have been granted a quite different term.

Consequently, the Office sets the term for each licence, using the same criteria of innovative capacity as it does for original grants. Any portion of the original term which has already expired before the licence commences, is deducted from the licensee's term.

It should be noted that if licences are granted to firms of lower "innovative capability" than that of the Patentee, it is quite possible that their terms will extend beyond the term of the original Innovation Patent grant. This is especially likely if licences are granted early in the Patent's life.

Kronz's proposals for licensing form an elegant counterpart to those for a variable term of monopoly; indeed, they may be regarded as essential if his variable term arrangements in practice are to be as flexible and sophisticated as they could be. Among the more obvious benefits of carrying the variable term to its logical conclusion in licensing provisions are:–

– Licensing would be much more "open", in the sense that a firm which can offer another a licence in return, might now face competition from a firm which can offer a long "term" in return. Under the classical Patent system, "cross-licensing", leading to the development of Patent pools and the like, reflects just the absence of licensing provisions of the type Kronz advocates.
– The problems of "know-how", "show-how" and "show why" which have caused such trouble in relation to Patent licences under the classical system, would be eliminated, since what is protected and disclosed is the actual commercial object, in all its working detail. Licences under the Kronz system, would therefore necessarily be comprehensive.
– Compulsory licensing is a procedure which was devised in an attempt to cope with the rigidities of the classical Patent system, which pays no attention to the innovative capacity of applicants. Since concern with this is central to Kronz's licensing system, the need for *compulsory* licensing correspondingly disappears.

INTERNATIONAL ASPECTS

The "Innovation Act" in Kronz's system is a combination of technical preparation and a commercial step. It is logical, therefore, for him to permit innovation protection to be obtained in respect of an imported product. If the technical preparation has been done abroad, importing the product into a country which grants innovation patents then constitutes the commercial step which completes the innovation act. A second corollary of Kronz's principle is that no infringement of an innovation patent takes place simply by manufacturing for export to a country which does not grant innovation patents. In such a case, there is only what he calls "latent infringement" in the country with innovation protection, since the second (commercial) step is accomplished abroad.

Kronz then fits his system into the rules of the Paris Convention by providing that no innovation acts performed in a second Convention country within a year of the provisional grant of an innovation patent in a first Convention country, can play any part in preventing the grant of an innovation patent in the second country, to the holder of the provisional grant in the first country. Others may also get an innovation patent for a similar (not identical) innovation, in the second country, but the scope of their claims will be limited by the "option" for protection held by the "first country" innovator for his Convention priority year.

In calculating the term in the second country, only investment made in that country will count for the term of protection. Consequently, although an innovation patent may be obtainable merely by importing a product (since "import" is the commercial step which, in the absence of anticipation, completes the innovation act) the term of protection may be short.

This is not just because only a small amount of investment (for transport, stockholding and marketing) is likely to be involved; it is also because the entry of a product into international trade can be argued to be evidence of its movement out of the innovatory and into the diffusion phase. Groups of countries (such as the EEC) may form a single "domestic" area for innovation patent purposes in the Kronz system.

RELATIONS BETWEEN CLASSICAL PATENT SYSTEM AND KRONZ SYSTEM

Kronz sees his system as capable of supplementing or replacing the classical patent system. He would regard replacement as no more than bringing the patent system, as this has actually developed, back to its original usefulness.

As long as both systems are in existence together, it can easily be envisaged that the subject matter of innovation protection could infringe a classical Patent, whether or not this Patent is exploited.

Kronz expects that some such cases will be resolved by the traditional doctrine in the Patent field, of "prior internal use". These would be the cases where the innovator had begun his work before the Patentee filed his application.

For other cases, he distinguishes according to whether or not the Patent is exploited. If it is not, he would use the provision which is widespread in Patent law, that failure to "use" the Patent within three years of grant, causes loss of rights. Where overlapping innovation protection is granted, Kronz would reduce this three year period to one year, after which the Patentee would have to issue a disclaimer of rights in respect of the specific innovation object.

If the Patent is exploited, Kronz accepts that innovation protection is not necessary (since there has been innovation). He would, however, modify the protection period of the patentee by reducing it from the normal Patent term of 18-20 years to whatever the Patentee would have been entitled to, if he had been an applicant for an Innovation Patent.

INNOVATION "QUALITY MARK"

This could be a representation, or designation (including a Trade Mark) combined with the number of the Innovation Patent. Its function is twofold: Firstly, to testify to the existence of a genuine innovation, in its innovation phase. For this reason, it is not permitted to be used in the diffusion phase, that is, it expires with the innovation patent.

According to Kronz, this "quality mark" is analogous to the name which has to be registered under Article 13 of the Convention for the Protection of New Varieties of Plants (the UPOV Convention). It could be used with advantage when computer software or genetic innovations are protected by an innovation patent.

Secondly, it signals the existence of "know-how" in the innovation object, since the latter is described by Kronz as "materialised know-how in a specific embodiment". It is an essential part of the Kronz system that the Specification of the innovation object at the time of grant must be absolutely complete, to the extent that anyone could reproduce and operate it without difficulty once the innovation patent had expired.

This is by no means the case with Specifications in the classical Patent system, and what is most frequently missing is the element of "know-how". The quality mark emphasises that the difference between innovation protection and invention protection includes the compulsory provision of "know-how" in the former.

INCORPORATION OF "DESIGN CLAIM"

Kronz provides for a claim covering the aesthetical features of the innovation object, to be included in the application. The resulting design protection can be separated from the actual innovation protection, and may outlive it. This contrasts with the "Innovation Mark" which is extinguished with the innovation protection.

CHAPTER III

THE INNOVATION WARRANT

INTRODUCTION

Like Kronz, Kingston advances proposals for direct protection of innovation. The fact that these differ from Kronz's suggestions in some aspects, only confirms that the principle of Patenting is capable of being applied in ways that go significantly beyond the scope of classical Patents.

Some of these proposals were advanced in Kingston's 1969 Paper, "Invention and Monopoly". At that time, they were seen as necessary steps in reform of the existing Patent system. When the ideas were later developed and extended, their central element was called an "Investment Patent". By then, the advantages of a new system which would be an addition to classical Patents, leaving the arrangements for these untouched, but also administered by Patent Offices, had been grasped. Still later, the positive benefits of having direct protection of innovation administered by some public authority other than the Patent Offices, made it necessary to find a name which would not impede radical thinking about how the proposals might be put into effect.

IMPORTANCE OF NAME

The object is to ensure that those who innovate can obtain a protection which is appropriate to the above-average risk which attaches to their investment. "Innovation Safeguard" would be an accurate description, if not an inspiring one. A far better candidate is "Innovation Warrant", because the original meaning of "warrant", coming from the old High German root "Gewähren" was *guarantee*. Reversion to this meaning in the present case would therefore cause no difficulty in the languages of any of the countries likely to develop the proposed new system.

Of the dozen senses of "warrant" in the Oxford English Dictionary, the first is "protector" and the second "safeguard". Another reference in the Dictionary which is interesting in the present context is to land grants in Colonial America, where both Patents and Warrants had a role. All land was originally granted by Patent from the Crown, and a Warrant established the right to such a grant of land, on the part of a veteran as reward for his military service.

If for "Patent" is read, "right to protection in ownership of designated property", and "making a risky investment in innovation" is equated with "military service", then a Warrant as giving a right to a monopoly reward for risky investment, which benefits the public in some way, is a good modern analogy.

The main advantage of using the term "Innovation Warrant" is that vested interests and inertia would have less power to stifle the emergence of the new kind of protection. If a Patent Office, or the Government department which is responsible for it, does not wish to extend the Office's scope, progress need not then be blocked, just because the word "Patent" is in the title. Innovation Warrants could issue from another source – which could be in a completely different Department. On the other hand, a dynamic institution in any country might turn itself into a Patent, Trade Mark and *Warrant* Office.

INNOVATION OFFICES

In practice, this is unlikely, and it is assumed in this Chapter that any arrangements for direct protection of innovation would eventually result from an initiative arising outside the Patent Office; that it would require a new and independent institution (called here, the Innovation Office) and that both Patent and Innovation Offices would run in parallel, one protecting invention, and the other protecting innovation. The existing Patent system, therefore, would remain completely untouched by the introduction of Warrants. It is believed, moreover, that many of the criticisms currently levelled by users and opponents of Patents would quickly cease to be justified, as a result of rivalry between the two institutions. Some of the shortcomings of the Patent system are intrinsic to it (such as the fact that it only protects innovation indirectly) but others have arisen – and certainly have persisted – only because the Patent Office in every country is a monopoly. A degree of competition could work wonders of improvement on it in a very short time.

There is, however, one group of Patent Offices whose members might see their salvation in turning themselves into Innovation Offices. These are the Offices of those EEC member-countries which have embraced the intentions of the European Patent Convention. They are already suffering from the rapid expansion in usage of the European Patent. The work of administering Innovation Warrants could more than compensate them for the work-load lost in this way, and ensure them a future. Warrants would also mean an expanded role for Patent Agents.

PROTECTING INVESTMENT

Direct protection of innovation by Warrants would be achieved by making the subject matter, not an idea, but *investment* to turn an idea into concrete reality. Warrants are an extension of the principle of Patenting, because they are an exchange of a monopoly, granted by the State for something the State wants to encourage. In this case, what the State wants is investment at above-average risk, in innovation.

Not every kind of investment can be protected by a Warrant, but only that kind which is concerned with getting *new* things done, that is, where new information is generated, and has to have protection. However, *anything* new can be protected, as long as it can be the subject of investment, which means anything that can be bought and sold. The restrictions which Patent Offices have placed upon their own operations, for example, by refusing to give protection to computer software or to methods of doing business, would have no place in the Warrant system. If money can change hands in connection with any of these things, investment to make money out of them is possible; if they are new, that investment requires protection, and the Warrant provides it, by means of a monopoly.

WHY A MONOPOLY?

Why is protection needed, and why should it be given by this particular means? There are two reasons for its necessity. The first of these is that investment in anything new is made under a correspondingly high degree of uncertainty, and is thus risky. Secondly, any investment of this kind is to some extent (often substantial) an exercise in generating new information. Once the object has been achieved, and a product put on the market, the information embodied in it is accessible to others who have not paid for it to be brought into existence. These therefore gain doubly, since by no means every attempt at innovation is successful, and they also avoid sharing in the losses of such attempts. The position is therefore not even a "neutral" one, if there is no protection for investment in innovation – there is an actual bias, founded on rationality, *against* such investment.

The granting of a monopoly, in accordance with the principle of Patenting, is capable of both protecting the new information contained in the embodiment, and enabling an above-average return to be earned on a high-risk investment. Clearly, there must be an appropriate correspondence of the monopoly granted to the objectives it is desired to achieve. A monopoly can be considered analogous to a container, and its term can be taken to be how long it takes to fill with water (money). If the size of the container is such that, when full, it would not hold enough to justify the investment in the eyes of a potential innovator, the investment will not be made. A container may also be leaky. If the outflow from the leaks is less than the inflow, the container will eventually fill, although it will take longer.

In one sense of the analogy, therefore, incompleteness of the monopoly can be compensated for by increasing its term. But the figures for returns to investment in innovation actually obtainable, suggest that in many cases the container is leaking so badly that money is flowing out as fast as it is coming in. No length of term can make up for the deficiencies of a so-called "monopoly" that "leaks" like a sieve. There is also an important way in which the analogy breaks down. A container full of liquid is the same whether it has taken a short or long time to fill. An innovation which is protected by a monopoly is not indifferent to time, for two reasons. Firstly, return on investment which comes in quickly, is worth more than if it is slow; secondly, the slower the return is in coming, the more likely it is that it will be stopped completely by the advent of a new product which supersedes the earlier one. The analogy may perhaps be extended to include the notion that the container is rusting through, and it may become utterly useless long before it has ever filled up, if the early rate of flow is not fast enough.

QUALITY OF PROTECTION

For these reasons, it is a fundamental premise of the Warrant system, that while it is highly desirable to have a variable term for the monopoly grant, to match it to the risk undertaken, it is impossible to make any sensible decisions about any actual term, unless the *quality* of the protection is taken into consideration also. Warrants as proposed, therefore, involve a considerable degree of effort to match the reward to the risk undertaken in an innovatory investment, not just by the length of the monopoly period, but also by how effective the protection is during it. The more perfect the protection is, the shorter Warrant terms can be, and at the limit they could be very short indeed. The Warrant system therefore puts quality of protection first in its list of objectives, since if this can be achieved, much else follows.

NOVELTY CRITERION

In the Warrant system, the definition of novelty is simple and precise, eliminating all abstract argument of the "inventive step" type which has caused so much trouble with Patents. It depends upon the answer to one question: "Is the subject-matter of the Warrant application available for purchase now in the ordinary course of trade?" If the answer to this question is "No", then investment to make the answer into "Yes" is entitled to the monopoly protection of a Warrant.

The range of protection potentially available is thus as wide as the range of Commerce. Anything that is or that could be bought and sold, comes within the scope of the system, which consequently extends far beyond technology. Incremental innovation, on which the Patent system has turned its back through its concern with "non-obviousness", fits easily into the Warrant system. If, for example, a product of a general type is available in the ordinary course of trade, but such a product with a particular new feature is not, then investment to put the product *which has the new feature* on the market is entitled to a Warrant, which would protect the new feature only.

At the outset, "available in the ordinary course of trade" can be expected to refer only to the national market of the country granting the Warrant. It should also mean that some genuine attempt to develop a market for the goods in question is already being made. There should be no question, in a Warrant system, of any device comparable to "token working" in Patents, where a provision of the Law can be satisfied by annual importation of a trivial quantity of the goods, or even by advertising an offer to licence the Patent. Ignoring all evidence which does not originate within the country also makes the task of the Innovation Office as easy as possible, since all information necessary for its preliminary screening will be readily to hand. This also applies to the information needed by objectors for opposition proceedings, which can therefore be completed quickly. Both these aspects of the assessment for novelty will be discussed further below.

It could be envisaged that once a group of countries with a roughly common technological level had gained experience with the Warrant system, they might want to move towards integrating their arrangements for granting this type of protection. Member-states of the EEC, for example, might extend their definition of novelty to "not available in the ordinary course of trade within the Community's boundary". With the rapid growth of data bases and computerised searching, national Innovation Offices should be able to cope with such a widening of the relevant area without any need to extend the length of time required for preliminary screening. The general growth of

Community-wide business links should also assist and accelerate the collection
of data for opposition proceedings, so that the period for these would not need
to be lengthened either.

INTERNATIONAL ARRANGEMENTS

Although such special arrangements, either bi-lateral or more general, may
arise at a later stage specifically for Innovation Warrants, these could certainly
be treated initially as Patents for the purpose of various international Agreements. The most important of these are the Paris Convention and the General
Agreement on Tariffs and Trade. Assimilation of Warrants to the provisions
of the Paris Convention, for example, would simply mean that a country
operating a Warrant system, would treat foreigners, if they are citizens of a
Convention country, in exactly the same way as its own citizens. That is,
foreigners would be just as much entitled to obtain a Warrant, but would, of
course, have to make the appropriate investment to put the new product, or
product with the new feature, on the market, and this investment would have
to be made in the country which grants the Warrant.

A firm might be expected to apply for a Warrant first in its own country.
Until the local Innovation Office concludes that there is a *prima facie* case for
novelty, the firm's information remains confidential to it, and it can decide to
apply for Warrants in other countries operating the system, but only if it
wishes to make an investment to produce there. If it does not do so, its
proposals will become public at the commencement of the opposition period,
and firms in those other countries might then apply to their Innovation Offices
for Warrants. The logic of the system is therefore that innovative activity will
tend to be widely diffused during the periods of Warrant monopoly. The fact
that mergers and rationalisation may be expected to take place on expiry of
the Warrant terms, strengthens the case for having Warrants incontestable and
fully protected for as long as they do last, and for having Warrant terms
sufficiently long to enable an investor in a risky project earn an appropriate
reward.

In attempting to achieve the Warrant's objectives, what must be avoided
above all is repeating the errors of the classical Patent system. The blunt truth
about a Patent monopoly is that its value as a deterrent to infringement is a
function of the resources available to its owner to prosecute an infringer
through the legal system. "The Courts", it has been said, "are open to each
and every citizen – like the doors of the Ritz hotel". In practice, private
inventors and even medium-sized firms do not have the resources to protect

their Patents, even if we assume these to be of a kind that would be held to be valid and infringed in a Court action. The inescapable consequence is that it makes no difference what term of monopoly is specified on the Patent document, if that Patent has issued to an individual who would have to spend his own money in defending it. To a potential infringer, it might as well state "Zero years", since he will not be prevented in practice from using the Patentee's invention, nor need he expect to suffer in any way from doing so. At the other end of the scale is the Patent owned by a Siemens or by a Du Pont. In such cases, the Patent term means what it says, because a potential infringer knows that no shortage of funds will ever be a factor in a decision about taking him to Court, with the risk of having to pay heavy damages and costs.

Infringement would also break the rules of the "cross-licensing" game which only large firms can play, and the results of this could be even more serious than damages awarded by a Court, for the firm's future. Indeed, it is likely enough that many Patents which would be ruled invalid if they were tested legally, have the effective strength of valid Patents, simply because no one is prepared to take the risk of infringing in the face of powerful resources for defending them. Thus, the 20-year term of a European Patent does not have a single meaning for all Patentees; its meaning and value vary over a range from zero to very great, according to the resources of the Patentee.

PUBLIC ENFORCEMENT

It is therefore proposed for Innovation Warrants, that the "quality" of the monopoly grant should be rendered completely independent of the Warrant holder, by making the enforcement of the Warrant the business of the Authority which has granted it. Apart from the advantages of removing the burden of protecting his monopoly from the Warrant holder, there is a particular logic to this approach. It is that a Warrant is granted by the State for a specific purpose, which is the encouragement of innovatory investment, this being considered to be for the common good. Consequently, an attempt to infringe a Warrant is not just an act which is actually or potentially damaging to a Warrant-holder; it is also an attack on the economic policy of the State. There is an element of lèse-majesté about it.

Once the principle is accepted that the State not only grants monopolies but enforces what it grants, there are several ways in which protection might be administered. At one extreme, infringement could be made a crime, as was done with the activity of "passing-off" one's own products in the place of another's in a British Act of 1862. In this case, the crime would presumably be that of "attempting to diminish the value of a monopoly granted by the State for the fulfilment of a public purpose", and would carry appropriate penalties, such as fines and/or imprisonment. For Innovation Warrants, the phrase "other than by innovation" should probably be added to this definition of the crime, since it would be the very opposite of the objective of the Warrant system to stifle development. It would therefore be desirable to state explicitly in the Law, that any Warrant-holder who is acting within the terms of his Warrant, shall be safe from all charges of damaging another Warrant-holder's monopoly.

At the other extreme, infringement of a Warrant could remain a Tort, as Patent infringement is at present. In this case, lifting the burden of protecting his monopoly from the Warrant holder would be through automatic availability of generous legal aid. That is, a Warrant-holder would engage his own lawyers and finance all the costs of taking the alleged infringer to Court, but would then be able to recoup all or part of his costs, when these are not awarded against the infringer, from the Innovation Office. There could be an advantage in varying the proportion of recoupment from, say, 50% in the case of substantial firms, to 100% in that of the modest ones, to achieve the objective of standardizing the quality of the Warrant monopoly, and rendering it independent of the resources of the Warrant-holder.

Between the two limits, it could be envisaged that the Innovation Office itself might prosecute infringers in the Civil Courts on behalf of the Warrant-holder. Most, if not all legal regimes have provision for Law Officers to proceed in cases where an individual is unable to do so for any reason. As a deterrent, the Law should provide that damages awarded should be exemplary, possibly on the model of the United States "treble damages" suit, which imposes a significant penalty on the infringer, as well as compensating his victim.

ACTIVE PROTECTION

There are several reasons why this intermediate position is to be preferred. Making infringement a crime would be a drastic step, for which the precedents in the industrial property field are, to say the least, obscure. The 1862 British Act was a step on the way towards full Trade Mark registration, and legal historians are not even sure whether anyone was ever punished under it. The most powerful argument against the "Tort with legal aid" approach is that it distracts innovators from their proper task. The whole purpose of the Warrant system is to direct creative energy towards innovative work. To win a legal battle requires a substantial drain on the same, limited pool of energy, with the inevitable result that innovation suffers and the broad purpose of Warrants is frustrated.

The intermediate position is supported by well-established arrangements in other types of property. A firm's investment at high risk in the generation of new information embodied in its products, may represent a far greater part of its total productive assets than its buildings or plant. It certainly is so, compared with whatever part of its current assets it holds in the form of cash in its factory. Thieves who break in and steal the money will be sought by the police and may be charged and convicted to a prison term; if the cash is taken by subterfuge by an employee of the firm, the Fraud Squad is available to be called in; both types of criminal may find themselves in prison in neighbouring cells to an arsonist who has burnt down a factory; in the unlikely event of piracy of an offshore oil rig, the Navy and/or Coastguard could be expected to provide policing; if an industry is protected by a tariff barrier, every Customs Officer has a sort of property-protection function. In the light of such parallels, it is indeed difficult to see why that part of a firm's assets that arise from its efforts at innovation, should have to be protected by the firm itself, and be denied any policing by the State. Active policing of this type, is, therefore, a fundamental part of the Warrant system as proposed.

INCONTESTABLE

Another aspect of the Warrant grant is that it would be incontestable, except on the ground that it had been obtained by fraud. This feature is also to be found in the Kronz system, and its necessity is so obvious that it requires little elaboration. Who would invest in exploitation of mineral "rights" if their geographical limits could be changed at any time by the emergence of previously unknown facts about the geological configuration of the territory? Yet this is not too far from the position of any present Patentee, who can see his Patent ruled to be invalid on the basis of a newly discovered document from the other side of the world, in an obscure language of which neither he nor any of his immediate competitors know a word. The Warrant system will allow ample opportunity in opposition proceedings, for genuine objections to be voiced, but once a grant is made, investment can be based upon it with complete confidence.

The fact that a Warrant is incontestable, and that the monopoly which it grants will be enforced by the State, so that the Warrant-holder is relieved of the burden of protecting it, places the question of length of the monopoly term in a completely different light. For example, a Warrant can provide a basis for rational investment on the part of a small firm, or even an individual, which a Patent could never do, no matter how long its term might be. Even for a very large firm, incontestability combined with freedom from litigation would make very much shorter terms than those of Patents at present, attractive as investment opportunities. Warrant terms can therefore be approximated to actual product life cycles. The strength of the monopoly commends a Warrant to its holder; its short term commends it to the holder's competitors.

COMPONENTS OF RISK

However, for the Warrant system, this is only the beginning of the process of matching the monopoly granted to the risk undertaken. Money can only be rationally invested in innovation if it can be anticipated that the return will be some multiple of what has been staked. Simply getting back the stake plus as much again, still less earning some "preference-share type" of percentage annually, cannot justify the risks which attach to investment in anything which deserves to be called innovation. Moreover, the multiple varies with the risk, of which at least two components can be easily identified. Firstly, there is the risk inherent in a project, almost irrespective of who undertakes it. The first attempt at making commercial products out of a genuinely new scientific advance, is an extremely hazardous process, as the record of biotechnology illustrates. New applications of well-understood technology carry a much lower level of risk, while the lowest level of all can be presumed to apply to incremental improvements to established products.

It is not, of course, strictly true to say that these different levels of risk are altogether independent of whatever firm might tackle the problem. For example, a project with the highest level of risk might pose *altogether insuperable* obstacles to all but a few firms in the world which have an exceptionally suitable combination of resources. At the same time, it is not inaccurate to refer to this type of risk as "project-related". There is a second type of risk which is "firm-related". This arises from the deployment of the firm's resources, of whatever kind, into any high-risk area. The level of risk depends upon how serious for a firm total failure would be, and this in turn can be held to be a factor of the size of the investment in relation to the resources available to the firm. Even in strictly financial terms, what might be handled from the petty cash of one of the largest multi-nationals could represent a suicidal level of investment in innovation for a very small firm. For the purposes of innovation, too, resources cannot simply be measured in money. Two firms which look identical to an Accountant, may be poles apart when it comes to the human and psychical resources needed to meet an innovatory challenge.

The actual risk as it is envisaged by the management of any firm which proposes to invest in innovation, therefore, is a combination of the project-related risk and the risk that is specific to the firm itself. The "multiple" of the investment that will be looked for as adequate return will consequently differ in firms with different kinds and amounts of resources, even if they are approaching the same project.

RISK "CATEGORIES"

In the abstract, therefore, the ideal solution to the Warrant "term" problem would be for each individual Warrant to remain in force until the firm had received the multiple appropriate to the combined firm-related and project-related risk undertaken, Measuring and monitoring costs render any "individual" approach impractical. Most of the advantages, however, could be obtained by accepting the principle of the "multiple", establishing *categories* based upon empirical research into risks and rewards in innovation, and expressing these categories in terms of time.

The following simple example will illustrate this proposal:–

Assign probabilities of success of 0.5, 0.3 and 0.2 to incremental innovation, to technology-transfer-type innovation, and to radical innovation, respectively.

Assign similar probabilities to three levels of firm-related risk, according to the relationship between the investment proposed for the innovation, and the firm's nett assets according to its balance sheet.

Then, the combined risk actually undertaken is measurable by multiplying these individual probabilities, which gives the following matrix:–

TABLE 1
COMBINED RISK FACTORS (PROBABILITY OF SUCCESS)

FIRM-RELATED RISK	PROJECT-RELATED RISK		
	Incremental	Tech-transfer	Radical
Low	0.25	0.15	0.1
Medium	0.15	0.09	0.06
High	0.1	0.06	0.04

Then, if the length of the monopoly term is to be the inverse of risk, in order to induce firms to innovate, for each category the number of years will be the reciprocal of these probabilities:–

TABLE 2
NUMBER OF YEARS' MONOPOLY

FIRM-RELATED RISK	TYPE OF INNOVATION		
	Incremental	Tech-transfer	Radical
Low	4	6.7	10
Medium	6.7	11	16.7
High	10	16.7	(25)

This indicates that a large firm carrying through a fairly routine incremental improvement to one of its products would have a monopoly of sales of products with that particular improvement for four years. A similar type of firm, addressing itself to a radical innovation from a base of substantial resources, only a small part of which would be committed to this project, would have a 10-year term for its Warrant. The other end of the scale would reflect a small firm, throwing much of its resources into a "make or break" effort to achieve some radical breakthrough, which would have a 25-year term in support of its commitment. Since the Innovation Office should not encourage the taking of risks of this extreme kind, the longest term to be offered in practice would be 16.7 years.

EFFECT ON TERMS

It will be observed that Warrant terms in this example, are generally shorter than Patent terms are at present. They would, nevertheless, give much more protection, and be a far greater incentive to invest, because they would be incontestable and because the Warrant-holders would not have to protect them – this would be done by the Innovation Office. Furthermore, Warrants would be offering protection where none is given by Patents. The 4 or 6.7 year terms for incremental innovations in the examples given, would presumably apply to improvements which would not qualify for Patent protection on the ground of "obviousness to one skilled in the Art".

The terms quoted in the example, of course, are simply for the purpose of illustration. In actual practice, each of the two types of risk could be further sub-divided, and additional types might be added, to increase the sophistication of the measurement. Also, the Innovation Office would constantly be carrying out empirical investigations, to improve statistical assessment of risk in its various categories, and the perception of these risks by business men. In a sense, this work would institutionalize the pioneering efforts to measure returns from investment in innovation already made by some economists.

Consequently, even with a system as primitive as that suggested, a potential investor in innovation could indeed look forward to obtaining a return which would be a multiple of his investment, in the event of technical success. Inevitably, at the outset there would be cases where the multiple actually received would fall below that envisaged by the authorities as the norm for a particular category, and there would also be cases where the investor would do much better than this norm. As the Innovation Office accumulated data and experience, it could be expected that its categories and Warrant terms would become progressively more effective in achieving the objective of matching the monopoly term to the risk undertaken.

NO ADMINISTRATIVE DISCRETION

There is a further important advantage of using categories of risk, as the means of establishing the term of a Warrant, which is that it eliminates administrative discretion on the part of the officials of the Innovation Office. If the system is working properly, a Warrant will be worth a relatively large amount of money, and a decision as to whether or not a firm is to be granted a Warrant, if left to a single individual, would place a heavy burden of responsibility on him. It would be considerably heavier than on a Patent Examiner, for example, since the relationship of substantial investment to a Warrant is, by definition, much more immediate than it is to a Patent. It would not be unfair to claim that the level of responsibility involved would be of the same order as that undertaken by judges. But of course, Innovation Office staff would neither be trained nor paid as judges are, nor could they have the same status as judges, any more than Patent Examiners. If any attempt were made to ask them to accept a kind of responsibility which is inappropriate to their conditions of employment, they could be expected to re-define the task so as to bring it into conformity with these conditions. This could only mean that the Warrant grants would be rendered weaker than they ought to be, with consequently reduced investment in innovation.

ANONYMOUS EVALUATION

It is difficult to see how such an outcome could be avoided if an attempt were made to establish the Warrant system on a basis of individually-assessed "multiples", even if Committee-decisions replaced individual ones. These difficulties disappear if categories are used, as suggested above. Project-related risks can be calculated in the abstract. Even if a proposal arrives for innovation in a field for which the Innovation Office has no previous data, the assessment of the difficulties to be faced, and the likelihood of overcoming them, could be done without any need to identify the applicant firm to those working on the assessment. In a parallel way, firm-related risks could be established without involving discretion on the part of any official. The amount a firm proposes to invest in the innovation forms part of its application, its resources in Accounting terms are available from its annual audit; factors which could modify the simple relationship of the two figures could be developed on a regular basis by the research department of the Innovation Office, and made available to all applicants. These might include whether the proposed innovatory investment is in the applicant's main field of activity, or whether it will involve any element of learning a new business; whether he has already-established R and D and/or experimental workshop facilities which are appropriate to the innovation proposed; whether he has attempted innovation in this or other fields before, and with what results; whether or not the project can be financed entirely from internally generated funds; the availability of taxation against which losses may be written off (i.e. the extent to which the applicant's risk is reduced by the partial, involuntary financing of the project by the Government); and the effect on the firm of total failure of this attempt to innovate.

With suitable weightings for such factors available, and constantly being up-dated by the Innovation Office's empirical researches, it should be possible for a firm to know, even before it makes an application, what its firm-related risk-factor is likely to be. No discretion on the part of Innovation Office officials is involved. Since the final category which determines the length of the Warrant term is arrived at simply by multiplying the firm-related and project-related probability figures, no official discretion is involved here either. All relevant figures would be available to interested parties during opposition proceedings. During these, if a firm considered that the term proposed by the Office is too short, it could present its arguments on the point, and these could be contested by its competitors. These could also give reasons why a proposed term is too long. It would be a condition of acceptance of a Warrant that all reasonably accessible information about the financial aspects of the innovation throughout its life would be divulged in confidence to the Innovation Office, to assist its work of measuring the risks and returns of investment in innovation with ever-increasing precision.

"COMMERCIAL EQUIVALENCE"

Some further attention must now be paid to the role of the Innovation Office as protector of the Warrants it grants. The remit of the Office as a whole would be to generate conditions which are conducive to investment in innovation, and its Legal division's part in fulfilling this, must be to take all necessary steps to ensure that each Warrant-holder's monopoly is effective. He has fulfilled his part of the bargain with the State by making his investment; the State must now fulfil its part (and, incidentally, stand on its own dignity) by protecting his Warrant.

In doing this, it will be necessary to look well beyond the doctrine of technical equivalence which applies in the Patent system. If monopoly is to mean what it says, the holder of the monopoly power must be safe from intrusions of any kind which diminish the rent which he is thus enabled to extract from the economy. There are many ways in which a competitor might diminish a Warrant-holder's rent, other than by producing a product which is technically equivalent. Some of these ways benefit the competitor, but some damage the monopolist without doing the competitor any corresponding (or indeed any) good. It would therefore be necessary for the Innovation Office to develop a new doctrine of "Commercial equivalence". This is so important that it would have to be spelled out in the Law establishing the Warrant system, and it could be envisaged that a jurisprudence of "commercial equivalence" would also develop through decisions by the Courts over time.

The criterion would depend upon the answer to the question: "If the alleged infringer's product had not been available, would the Warrant-holder have made a particular sale?" Factors which would obviously have to be considered would be the determination of customers to buy some product of the kind, frustration of efforts to sell on the part of the Warrant-holder, evidence of educated buyers in the particular market, and the marketing objectives of the alleged infringer. Such protection is evidently far wider than anything granted by a Patent, but it is limited in two main ways. On one side, no product which was on the market at the time of the application for a Warrant, can be affected. On the other, no product which is itself the subject of an Innovation Warrant, can be held to infringe any other Warrant.

This is because in the Warrant system, "infringement" has the meaning "attempting to diminish the value of an Innovation Warrant, *other than by innovation*. Amongst the risks which any investor in innovation takes, is that his product will be rendered obsolete, and lose its market, by an even more advanced innovation. This risk, like all others, is taken into account in establishing the "categories" and monopoly terms. It must be foreseen, nevertheless, that the granting of a particular Innovation Warrant will occasionally undermine an earlier one.

TEMPERING RATE OF CHANGE

Although according to the strict logic of competition, this should be allowed to happen without interference, there could be a role for the Innovation Office in reducing the cost of rapid change, especially in human terms. Public acceptance of innovation will be all the more ready if the change is made less brutal in its effects on employees by arrangements to prepare for it in advance, and by tempering both its speed and its impact when it does come. When it becomes clear during opposition proceedings that innovation of the subject-matter of a new application would have serious implications for the survival of an earlier innovation that has Warrant protection, the Innovation Office might have power to issue the new Warrant with some special conditions attached. A possible condition might be that the new innovation would pay a royalty to the one which it displaces, the rationale being that every new development is made possible in practice only through the existence of earlier ones. This is not just because the new makes use of technical information generated by the old, which is always the case, even if it may not always be possible to trace the precise linkage.

Much more important may be what the earlier innovation has done to introduce a product with a particular feature to the public, and to build up a market for it. Another means by which this debt might be recognized is for the new innovation to be obliged to share in any retraining payments which have to be made to the workforce of the earlier innovation, as a result of its displacement by the newer one. In administering such a policy, the Innovation Office would have to steer a middle course between slowing down the pace of change, and allowing change to have social effects that may ultimately cause a rejection of the processes of change by public opinion. Its objective is to render successful innovation commonplace instead of exceptional, and it needs to carry public opinion along with it in its efforts towards this end.

SHORTER TERMS

Paradoxically, shorter monopoly terms could result from a policy of tempering the effects of change. This is because a major force against a positive decision to invest in innovation is uncertainty concerning the possible rapid emergence of yet another innovation which could make the earlier investment worthless. The countervailing force is the attraction of the potential profits from a long monopoly term, if no such other innovation does appear. The length of monopoly term to produce just enough attraction to overcome the inertia induced by this type of uncertainty, is less if an investor can foresee that all will not be lost if a better product comes on the market, since the new product would be forced, as a condition of its own Innovation Warrant monopoly, to mitigate some of the losses it imposes on the older one. The shorter that Warrant terms can be, while still achieving their end of "luring capital on to untried trails", to use Schumpeter's words, the more easily the entire Warrant system will obtain public acceptance and support.

It would be important that any such interference with the stern operation of the law of survival of the fittest, could be operated without requiring the use of discretion by officials of the issuing Office. Discretion would be demanded of them, for example, if they were asked to accept or reject arguments from existing Warrant-holders during the opposition stage of applications for new Warrants, that their interests would be adversely affected, and then to prescribe what payments the new innovator might have to make to the old one. Instead, their duty would be to devise appropriate rules in advance, to cover the situations envisaged. These rules would indicate to a prospective investor in a project which has been offered a Warrant, that success in obtaining a share of a market at the expense of another Warrant-holder, would involve him in additional costs. Such costs, weighted by the likelihood of incurring them, would then come into his range of calculations leading to a decision whether or not to accept the offer of a Warrant, and proceed with the project. To reduce or even just to quantify his risk in this matter, he might negotiate with the owner of the earlier Warrant. This might be expected to take place most frequently when the latter has a long term to run, such as where a relatively small firm is pioneering the commercialisation of a piece of fundamental research.

SPECIAL POLICY "FACTORS"

The Innovation Warrant could be an extremely flexible instrument of Science, Regional or other type of public policy. All that would be required is provision in the Law which establishes the system, for additional factors, capable of being prescribed by Statute or by Ordinance, to be applicable to the risk factors which are used by the Innovation Office in establishing monopoly terms. For example, it might be provided that investment in areas with high unemployment would have twice the monopoly term which would apply elsewhere in the country. Or, if it were considered desirable to direct innovative efforts towards, say, software development or genetic engineering, application of an additional factor of 1.5 would certainly stimulate investment in these, as compared with other areas. Not alone would such factors be costless to administer, but the investment which they would steer in particular directions, corresponding to public policy, would still be subject to the discipline of the market. The extra advantage from the special "factor" would only be gained if technical and commercial success were achieved. This is in marked and favourable contrast to present arrangements for Regional and similar policies, where the cost to the taxpayer is the same in the event of failure as it is for success.

OUTLINE PROCEDURES

A firm which wished to take advantage of the Warrant system would submit a proposal for investment to put something new on the market, to the Innovation Office. The innovation might be a completely new product, or it might be a product already on the market, but with some new feature. In the latter case, of course, the Warrant protection would only relate to the novel aspect. The application would have to specify the amount of investment estimated to be required to carry through the innovation, and a certificate from the firm's auditors as to the relationship of this amount to the firm's nett assets would also be needed.

The Office would then immediately carry out an initial screening process, to eliminate applications which can easily be shown to cover items already on the market. When the Office can demonstrate this from its own monitoring resources, the applicant will be advised accordingly. This should normally extinguish the application. There would, however, be a right of appeal to the Courts, at the applicant's own risk in respect of legal costs, from such a decision.

Where the Office's preliminary screening indicates that what is proposed is *prima facie* new to the market, its technical details will be published, and a period allowed for Third Party objections. Because a Warrant would be incontestable except on the ground that it had been obtained through fraud, monitoring of such published applications would be vastly more important to all firms than monitoring of Patent Specifications is now. This is a valuable feature of the Warrant system from the aspect of improving awareness of the state of the art amongst all types of businesses, and of stimulation of innovatory activity as a result. For this reason, efforts would be made from the establishment of the Warrant system, to develop automatic computerised monitoring arrangements, real-time and on-line, to bring new applications to the notice of interested parties as early as possible. The period for Third Party objections would include oral hearings if these could help to clarify issues.

After considering any new information made available in this way, the Office would reject certain applications. Such a rejection could also be appealed to the Courts at the applicant's own risk in respect of costs. Where there are no grounds for rejection, the Office will calculate both project-related and firm-related risks, combine them, and offer an option on a Warrant for the appropriate term to the applicant. As mentioned above, there could also be special "Policy Factors" to be applied, which would change the term according to the area of the proposed investment, or in accordance with other objectives of public policy. These factors would be publicly known, so the applicant would be able to choose between a number of practical possibilities for making his investment, such as geographical location etc. Exercising an option would commit the firm to make the investment proposed in its application and the Office to protect its monopoly throughout the term of the grant, both according to procedures which will now be explained in more detail.

OPTION PERIOD

When it has been accepted by the Innovation Office, after opposition proceedings, that the subject matter of a Warrant application is indeed "not available in the ordinary course of trade", it will offer a Warrant to the applicant. This is the first time that the project can be considered in full seriousness as an investment opportunity, since it is only now that a known monopoly will compensate for the risk involved. Some period of time must now be allowed for the applicant to make his detailed plans, arrange financing, and reach a final decision as to whether or not to make the necessary investment. Although the applicant should not be forced into a hasty decision, this period should equally be no longer than is strictly necessary, since during it competitors do not know how far their own activities may come to be circumscribed. (Since no Warrant has yet issued, they are not limited in any way throughout this option period).

Instead of a period of absolute duration, therefore, one related to the length of term of the offered Warrant is more appropriate. It can be envisaged that decisions and arrangements in respect of more radical advances and involving a relatively bigger commitment of a firm's resources, will take longer to decide upon finally, than incremental innovations. A period of one-twentieth of the term of the Warrant would give options in a range from two and a half to ten months on the basis of the matrix suggested earlier, which does not seem unreasonable. Evidence that the applicant has the money for the necessary investment available, would also be a condition of granting a Warrant (though not, of course, a condition of offering an option).

INVESTMENT PERIOD

The continuance of a Warrant in force is conditional on making an investment to carry through the innovation, and once again, this must be time-bound. Allowance must be made for changes in the investment plan as a result of learning since the application was originally made, both from the opposition proceedings and other sources. For the shorter monopoly terms, it might be generally regarded as reasonable that the investment should be completed within one-eighth of the term, if the Warrant is not to lapse. However, the whole area is proper subject-matter for continuing study by the Research Department of the Innovation Office, and it would not be surprising if the result of its work were to be stricter conditions in relation to investment in incremental innovation and more relaxed ones for the more radical types.

PREMIUM FOR RAPID INNOVATION

A firm which has already made all arrangements for a decision to be taken to make its investment immediately it is offered an option on a Warrant, effectively adds the option period to its monopoly term. The more rapidly it can make its subsequent investment and get its product on the market, the more of its monopoly term will be valuable in terms of protected revenue. The approach outlined therefore puts a premium both on speed of innovation, and on being as far advanced as possible at the time an application for a Warrant is made. The latter, of course, would have to be balanced by any potential applicant, against the danger that a competitor might apply for a Warrant first.

TWO-STAGE APPROACH

Also, for the longer Warrant terms, a two-stage approach to the making of the main bulk of the investment corresponds to economic reality, to risk as perceived by firms, and to the desirability of eliminating monopolies which are not being used effectively, from the Warrant system as early as possible.

Such a two-stage approach might grant a Warrant initially for, say, one-third of its term, conditional upon one-third of the investment being completed by the mid-point of this period. In the case of a 16.6 year term from the matrix suggested earlier, the monopoly would therefore be granted in the first instance for just over five and a half years, but would lapse if one-third of the investment specified in the application had not been made within two and three-quarter years. Assuming this condition is met, renewal for the balance of the term would be automatic if the firm had already completed its investment or commits itself formally to do so, within a further specified period – say, one-eighth of the total Warrant term, about two years in the example quoted. Again, failure to fulfil this commitment would cause the Warrant to lapse finally.

Naturally, the actual phases would be adjusted in practice according to the empirical findings of the Research Division of the Innovation Office, but it is clear that in principle the use of phases has advantages to all parties. The investor knows that his monopoly is certain for the prescribed term, subject only to his making the investment which is necessary to take commercial advantage of it. This investment is to be made in amounts, and at intervals, which conform to reasonable expectations of the growth of learning and experience of the product and of its market, By the time it is necessary, therefore, to make the major part of the investment so as to obtain automatic renewal of the Warrant for its second portion, this can be done on the basis of an appropriate amount and quality of information.

HANDLING FAILURES

From the standpoint of the Innovation Office, it is just as important that Warrants do not impede legitimate business activity on the part of non-Warrant-holders, as that Warrants make money for those who do hold them. Warrant monopolies held by firms which are not succeeding with the intended innovation, may also be preventing the same problem from being attacked by another firm. The fact that a Warrant confers really valuable protection should ensure that a market for Warrants would develop, and this could be relied upon to resolve many, if not indeed most such cases. Some of the remaining problems would be mitigated by the fact that from the offer of a Warrant to an applicant, there are four points during the longer terms of monopoly, when others may become free to exploit the Warrant's subject-area, if they wish to do so. The applicant may decline to take up his option; he may fail to complete the first instalment of investment within the prescribed period; he may decline to commit himself to the further investment which is a condition of automatic renewal of his Warrant; or he may fail to make this investment in time.

WHEN WARRANTS COULD LAPSE

The following table illustrates the approximate timing of all the points where decisions have to be made about actions on which continuation of a Warrant in force, could depend:–

It will be noted that for the shorter terms, investment must be completed within about one-sixth of the Warrant term, if the Warrant is not to lapse. For the longer terms, the completion of investment must be by roughly the half-way point of the Warrant term, but there are two points leading up to this, when a decision about parts of that investment have to be made. If there is no positive decision at these points, coupled with certification to the Innovation Office that the necessary funds are available, the Warrant lapses. In addition, if the corresponding proportion of the investment has not been made by the expiry of the prescribed further part of the Warrant term, the Warrant would also lapse.

TABLE 3
ELAPSED TIME (YEARS) TO VARIOUS DECISION POINTS FROM OFFER OF OPTION FOR A WARRANT.

WARRANT TERM	EXERCISE OPTION	ONE-THIRD OF INVESTMENT	COMMITMENT TO MAKE BALANCE	INVESTMENT COMPLETE
4	0.2	n.a.	n.a.	0.7
6.7	0.3	n.a.	n.a.	1.1
10	0.5	2.2	3.8	5.1
11	0.6	2.4	4.3	5.7
16.7	0.8	3.6	6.4	8.5
n.a. = Not applicable				

Of these four decision-points, it is the third, in relation to automatic renewal, which is likely to be most effective in changing the ownership of Warrant monopolies, so that they are put into the hands of those who can make the best use of them. By that time, as a result of investment of one-third of the total envisaged at the time of application, by the Warrant holder, he will have obtained quite a precise notion of the problems and potential profits of the innovation concerned. Watching his efforts from the outside, other firms, too, will have been refining their own estimates of the potential of that particular Warrant.

The Warrant holder may decide to provide the second tranche of investment from his own resources, or to raise further capital, or to sell out to a competitor who rates his own chances of success with the innovation more highly. Because the relevant amounts of money will be significant, he is most unlikely to look for renewal of his Warrant simply to prevent others from doing what he has so far failed to do himself. Renewal points could therefore be expected to be times of widespread re-adjustment of Warrant ownership, marked by effective clearing out of all but a very few Warrants that have potential which is not being exploited.

"SURRENDER AND RE-GRANT"

Another device which the Innovation Office might use in cases where altogether unexpected difficulties were encountered is "Surrender and Re-grant". An example of the sort of problem to which this could apply is recombinant DNA, where in many cases "scaling-up" to commercial quantities has been found to be impossible. Worse, genes which co-operate very well in laboratories, can become quite mutually antagonistic on a commercial scale. Since a Warrant would normally have been applied for, and awarded, on the basis of the "laboratory" knowledge in order to bring about commercialization, problems of this magnitude would be altogether beyond what any of the parties could have expected. In such exceptional circumstances, the Innovation Office might declare the field to be one to which its "Surrender and Re-grant" policy applied. This would be the equivalent of declaring it an innovation "disaster area". It would mean that without another examination for novelty, the Office would be willing to accept the surrender of an existing Warrant, and to grant a new one, to the same firm re-structured, or to a joint venture, or to a new firm (which presumably would have reached prior agreement with the Warrant-holder).

It could be envisaged that use of such a device would come into consideration most frequently around the time when automatic renewal approaches. A firm which has run into technical difficulties by then, and which is faced with the need to make an investment which is twice the size of what it has so far risked, is most likely to see and value the advantages of seeking help from another firm or firms at that time, so as to carry the innovation through to success. Such help would probably be technical as much as financial. This phasing also corresponds with the psychology of the innovator. Acting under uncertainty requires a particular kind of confidence which is possessed by innovators and not others, but this confidence also makes innovators unwilling to share management of their projects, especially at the outset. After a number of setbacks, however, when they have greater appreciation of the scale of the difficulties they face, they are less reluctant to accept help, even if this involves some management role for the helpers.

The policy of "Surrender and Re-grant" effectively enables a fresh start to be made, bringing new resources to the project, preserving what has already been achieved, and extending the time-scale realistically in accordance with experience of what is required. Such a policy, like all the others of the Innovation Office, would be constantly monitored by its Research Division to assess its effectiveness in contributing to the achievement of the objectives of the Office. Amongst these would be making the monopoly granted by a Warrant progressively more effective in commercial terms, and in matching rewards to perceived risks. This would be so that the maximum amount of investment in innovation could be brought about with ever shorter Warrant terms in every category.

TIMING OF OPERATIONS

There are strong arguments for having all decisions in Warrant cases made with the utmost speed, and the simplicity of the novelty criterion permits this. For preliminary screening, a few telephone calls to manufacturers and import agents may be all that is necessary for a ruling to issue within a month of the application. If the Specification of what is proposed is clear and detailed enough, it can be published immediately, in the "favourable" cases, and the period of opposition will then commence. Otherwise, this period will start as soon as the defects in the Specification have been remedied by the applicant. Three months should be adequate for opposition to be made, including oral hearings if necessary. A decision should issue within a further two months, making six months in all from the application date, if the applicant has done his part of the work properly.

STRUCTURE OF THE INNOVATION OFFICE

A typical Innovation Office might have four main divisions, as follows:–

DIRECTION

SECRETARIAT   RESEARCH   ENFORCEMENT   EXAMINATION

In contrast to a Patent Office, which is almost wholly devoted to examination of applications, an Innovation Office would have two other departments of substantially equal importance. These would be concerned with Research and Enforcement respectively.

RESEARCH DIVISION

This would have responsibility for carrying out continuous research into returns obtained from investment in innovation of every identifiable kind. The results would be used for establishing the most accurate possible probability figures as a basis for setting monopoly terms which would properly match reward to risk in the event of technical success. It would also have a function in the examination process by advising on the "project-related" risk of individual applications submitted to it anonymously by the Examination Division. It is desirable that firms should be able to obtain such rulings in advance of making a formal application, since the preparation of the latter involves effort which should only be undertaken if the term likely to be awarded, is long enough.

EXAMINATION DIVISION

This Division would carry out the preliminary screening of applications to establish whether or not there is a *prima facie* case from the aspect of novelty, for permitting the application to proceed. For this, it would have its own research facilities, which would endeavour to have as much up to date information as possible about what is "available in the ordinary course of trade". The Examination Division would arrange for the publication of the relevant portions of applications which are going forward, for the organization of opposition proceedings, for getting rulings from the Research Division on project-related risks, for obtaining the information needed for establishing the firm-related risk in any particular case, and for advising applicants of rejections and the Enforcement Division of approved applications.

ENFORCEMENT DIVISION

This Division would make offers of options for Warrants, issue the Warrants when options are taken up, and deal with all cases of alleged infringement, both from the point of view of the insult to the authority of the State as well as those of material damage to a Warrant holder, which would be involved in "any attempt, other than by innovation, to reduce the value of a monopoly granted for a public purpose by the Innovation Office".

This Division would also monitor commitments to invest and actual investments prescribed to keep Warrants in force, and administer the policy of "Surrender and Re-grant".

Acceptance of a Warrant would involve handing over to the Innovation Office, Power of Attorney to act in all ways necessary for protection of the holder's rights arising from it. The Enforcement Division would exercise these powers promptly against all reported infringers, obtain injunctions to minimise harm to the Warrant-holder's business until a case can be heard, and seek exemplary damages from the Court. Its policy would be to get such a reputation for swift and determined action that infringement would only be risked in the very rarest cases. This would be reinforced by developing public awareness of the "lèse-majesté" aspect of infringement.

MISCELLANEOUS FEATURES

There would be no renewal fees for Innovation Warrants. The innovating firm "renews" by its commitment to invest, or by actual investment.

If a firm which holds a Warrant falls under the control of another during the Warrant period, the Warrant term is adjusted accordingly. This prevents a small firm being used as a "front" for a large one, so as to obtain a longer Warrant term. Raising extra capital, without surrendering control, would not involve any change.

Prior internal use of an innovation offers no bar to the grant of a Warrant to another firm, since such use does not make the innovation "available in the ordinary course of trade". Such internal use, however, would not have to be ended as a result of the grant of a Warrant to another firm. A process would be protected by a Warrant through what it produces, which would have to be "new". The law establishing the system would specifically provide that a substantial reduction in price would count as novelty. Investment in a process to produce at the lower price, would therefore be protected by a Warrant. The monopoly would not apply, of course, to products offered for sale at a higher price than that specified, but would prevent competitors from selling at the lower price. "Significant" lowering of price would mean being able to go measurably further than the well-established rule of 25% reduction in costs for every doubling of cumulative output, would allow.

It is desirable that the highest possible degree of certainty should apply to each and every limit of any grant, so that all firms will know precisely what they may, or may not, do. In the U.S., it is possible to obtain a declaration of validity of a patent from a Court, although this provision is almost never used, because of the expense involved. It should be possible for either a Warrant-holder or anyone wishing to compete with him, to obtain a rapid "Expert Declaration" which would not have the same status as a Court ruling, but to which a Court would have to pay attention. In any infringement action, a Warrant-holder would be on stronger ground if such a Declaration had been in his favour, and vice versa. Similarly, if infringement is proved in a Court, the damages awarded would be greater if the infringer had acted, either without first seeking such a Declaration, or after one unfavourable to him had been made.

The creativity which is the core of an innovative firm is a rare and extremely scarce resource for the public good; it is important that this should not be diverted into unproductive channels such as litigation. Consequently, it is the Office's job to prevent this, by heading off litigation through its Declarations, and by carrying the burden of infringement litigation itself.

Infringement should be a matter of commercial equivalents, not just technical ones. This should take account of the reality that the truly innovative act, commercially speaking, may be the one of uncovering a hitherto unknown market.

SUMMARY

An Innovation Warrant would therefore extend the principle of Patenting so as to give direct monopoly protection for investment in anything not already on the market. The Innovation Office would enforce its grants, because infringement involves "lèse-majesté", as well as damage to the Warrant-holder. Above all, enforcement is the key to shorter terms. Terms of Warrants would vary, depending upon both project-related and firm-related risks, and all Innovation Office procedures would be subject to constant scrutiny and revision.

CHAPTER IV

ADVANTAGES OF PROTECTING INNOVATION DIRECTLY

The proposals outlined in the two previous chapters were developed completely independently of each other. One proposal comes from a legal and Patent Attorney background; the other from business and economics. The extent of their agreement is consequently quite remarkable, including, as it does:–
– Subject matter of protection to be innovation, not invention.
– Any economic object can be protected, not just technology.
– Variable term of monopoly grant.
– Money, not time, is explicitly recognized as the proper measure of the monopoly granted (as it is, of course, of any monopoly).
– Time is only used as a convenient surrogate for the money measure in a Grant.
– Examination relies heavily on Third Party involvement.
– Grant is incontestable except through "fraud on the Office".
– No renewal fees.
– Terms of grant can differ between regions of a country, or can apply over more than one country.
– System to be administered by an independent Authority.
There are four main differences:–
– The Innovation Patent requires that the "innovation object" should exist before protection is given, which means that the associated investment would have to be made first. However, such a Patent can be taken for granted if the "innovation object" is in fact put on the market. The Innovation Warrant offers protection before investment has to be made, but actual investment is the condition of keeping the monopoly in force.
– The Variable Term arrangements of the Innovation Patent are highly sophisticated as means of matching the length of monopoly exactly to the individual "innovative capacity" of the Patentee. The Innovation Warrant proposals set a high value on eliminating official discretion, and accept that the price to be paid for this is a set of terms which, whilst being generally effective in relating monopoly granted to risk undertaken, may not correspond to the "innovative capacity" so precisely in individual cases.
– Licensing of an Innovation Patent involves the granting Authority, which

will prescribe terms that take account of the "innovative capacity" of the licensee. No similar provisions apply to Innovation Warrants.

– Protection of the Innovation Patent grant is left to the Patentee, but the idea that the State should shoulder the burden of protecting the grants it makes, is fundamental to the Warrant system.

DIRECT PROTECTION NOW PRACTICABLE.

Direct protection of innovation emerges from these proposals as a practical possibility. A large part of the hitherto untapped potential of the principle of Patenting could be exploited by taking the best features of each of the two systems suggested. Moreover, social innovation is no exception to the rule that all innovation is a learning process, and it is certain that many further improvements would be found to be possible, once a system was in actual operation. Both sets of proposals, it should be noted, make use of facilities and techniques which were not available at the time that the classical Patent system turned away from protection of innovation directly, to protecting it indirectly by making invention its subject matter.

The Innovation Patent's use of market research and evaluation of the applicant's innovation potential in setting the term of monopoly to be granted; the Innovation Warrant's reliance on Accounting and uncertainty-measuring techniques for the same purpose; and the way modern communication systems and computerised searching are essential to the evaluation of novelty in both, may be quoted as examples. To the extent that any of these techniques are essential to it, direct protection of the types of innovation which came in with the Industrial Revolution has not been fully possible at any time before the present.

Given that it can be done now, however, it becomes worth while to evaluate the advantages which establishing it could bring about. These will be found to consist mainly of filling gaps which have been left by limiting the exploitation of the principle of Patenting to the protection of invention.

LOW PROFITABILITY OF INNOVATION

There is no shortage of evidence that in those areas where the Patent system is supposed to protect high-risk investment in Western countries, the profitability of innovation is low. Moreover, it appears to have been dropping in recent years. The academic research available is predominantly of United

States origin, but there are good reasons for thinking that European results would point in the same direction.

These empirical studies depend upon attempts to measure private and social returns to investment in innovation. Three of them, analysed by von Hippel, use Mansfield's definition:—

"In brief, Mansfield defines private returns as net pretax profit of the innovating company minus R & D costs and other innovation-related investment minus profits (such as profits from sales of products displaced by the innovative product) lost as a result of the innovation. Social returns are defined as private returns plus innovation-related benefits and costs incurred by all innovation users and firms which compete with the innovator plus the impact, if any, of the innovation on public goods such as water quality.'(1)

von Hippel points out that

"Mansfield's measure is conservative given the direction of his findings, because it excludes returns from investment in innovation gained by suppliers, independent inventors and others".

Now, since the function of any type of market power is precisely to capture the returns from investment in innovation for the innovator (i.e. to prevent them from "leaking away" to competitors or other "free-riders") the extent to which social returns exceed private returns reflects failure on the part of the protection system. As observed in Chapter I, no actual innovation actually depends upon the market power conferred by a Patent alone, and in fact the evolution of Patents has been in the direction of making them subordinate to other types of market power, especially that of Capability.

No empirical study, therefore, can be claimed to be more than a very crude measure of the contribution made by Patents only, to any difference recorded between private and social returns. It is just not possible to quantify the contribution made by any one of several types of market power in any individual case. On the other hand, empirical studies of innovation frequently use a definition of it which inclines towards the types of innovation where Patents tend to be used. For example, such studies would not include innovations protected mainly by Persuasive market power, in spite of the economic importance of these.

Some compilers of the Baruch Report on Innovation in the U.S. contrasted the alleged shortage of funds for technological innovation with a typical $18 million investment in a new cosmetic product, reported to them. However, they did not advert to the extent to which the contrast might be due to differences in the capacity of different types of market power to capture the returns to investment in innovation. What can be asserted confidently, nevertheless, is that if Patents plus other types of market power fail to prevent a substantial amount of "leakage" of returns, then the situation could only be worse – and probably far worse – if Patents alone carried the entire burden of protection.

RATE OF RETURN

In the first enquiry considered by von Hippel (that of a team led by Mansfield himself) the median social rate of return from 17 cases was 56%, and the median private rate, 25%. (2) In two other studies, each of 20 cases, the respective rates were 99% and 70% for social, and 27% and 36% for private returns.(3) These figures indicate that the aggregate protection mechanism is not working well. Such returns (which are before taxation) are hardly better than a firm might expect from completely modernising its plant, and by no means correspond to any level of risk which could compensate for the failure rate that is inherent in innovation. The corollary must be that Patents, as a component of the total protection mechanism, are even further away from achievement of what is supposed to be their objective, which is capturing all returns to investment in innovation for the innovator. von Hippel, in fact, concludes that:
"patent grant is not a very effective innovation benefit capture mechanism in most fields of technical endeavour, with pharmaceuticals and chemicals being the primary exceptions. Data show innovators do not rely much on patent protection, and gain little financial return from patents they attempt to licence." (4)
This conclusion from an authoritative survey of empirical evidence, is exactly in line with what would be expected from the analysis in Chapter I, of the way the Patent system moved from direct to indirect protection of innovation. It also confirms that the inadequacy of the classical Patent system is most marked in non-chemical innovation.

Poorly as the innovation protection arrangements were shown to work in the studies quoted by von Hippel, there is also evidence that the profitability of investment in innovation is actually falling. In real terms, company-financed R&D in the U.S. grew at 6.3% a year from 1960–61, remained on a plateau for two years, and then started to grow again, but at a much slower rate. Scherer has shown how spending on R&D reflects – with a time-lag – its profitability. One cause of the lower profitability could be a combination of increased competition (e.g. from abroad) with a reduced pool of innovation opportunities. But Scherer's work also confirms the earlier studies which showed a large divergence between social and private returns to investment in R&D.

LESS PROFITS FROM R&D

Reduced R&D spending could therefore also be attributed to an increasing divergence between social and private returns, meaning that the protection mechanism is functioning even less effectively than before. (5) Scherer's empirical work shows that the marginal productivity of R&D increased in 1973–78 over 1964–69. This lends support to the hypothesis that R&D retained its social productivity, but investors were capturing a smaller proportion of the returns to their investment in innovation. Hence they became dubious about the value of spending on R&D and cut it back. As will be seen below, this is perfectly consistent with the evolution of market power systems, not alone in the U.S., but in all Western countries, combined with the impact of Japan on international markets.

Whether or not the situation has actually been getting worse, it would be impossible to pretend that allowing Patents to evolve in the way they have done, has resulted in conditions that are conducive to investment in innovation. A shift to direct protection of innovation, however, would immediately transform these conditions for the better. Protection of any investment in innovation by either of the systems proposed, or by any combination of them, would evidently be immensely stronger than anything available from the existing Patent system. Investors could count upon higher private returns to their R & D, which, without affecting overall R&D productivity, would transform its profitability. This would lead to further expenditure on R& D, and thus to further social and private returns in the longer run. Such a strengthening of protection would be socially acceptable because of a feature which is common to both proposals, the variable term of the monopoly grant.

VARIABLE TERM

The disadvantage of the fixed term for a Patent has been adverted to from time to time, recently, for example, in the British Green Paper on Industrial Property and Innovation. (6) It has not been appreciated, however, that anything else is altogether incompatible with indirect protection for innovation, which is all that Patents for invention can give. As long as the subject matter of protection is the invention, i.e. only the concept or the "teaching", what reliable basis can there be for deciding on one term rather than another?

If "likely public utility" is the criterion, a Patent Specification is a poor basis for judgment. The history of early rejection of inventions which later became major innovations (Kettering and his self-starter, Whittle and his jet engine, Land and his camera) by experienced *firms*, is well documented. How could Patent Office examiners be better prophets? From the Patent Specification to a successful product on the market is a very long way, and who can tell what difficulties will be encountered on the journey? The Patent is valid if the Specification discloses something new that works, even if it does so badly. The market will only accept whatever works *well*.

At the grant stage, there is no evidence even as to the width of this gap, much less as to what will be involved in bridging it. Float glass is largely disclosed in a United States Patent of 1904. It became a practical reality a half-century later, but in Britain. How could the U.S. Examiner have made a sensible recommendation about an appropriate term? Similar problems arise if "complexity" or "inherent difficulty" are made the measure of the term. "Fulfilling a long-felt want" is hard enough to handle already as testimony to the existence of an "inventive step", and in any event involves evidence other than what is disclosed in the Specification. Could the term differ according to "conceptual elegance"? Scarcely. Beauty in design comes at the end, not at the start of the process. For all these reasons, it seems inescapable that any attempt to prescribe varying terms for Patents *for invention* would place an intolerable burden on the subjective judgment of Patent officials, and even then would not satisfy users of the Patent system.

The basic reality is that the length of the monopoly term is quite meaningless except in relation to *an identifiable innovator*. If he is to be able to act, others must be kept out of his market for a time, and this time must strike a balance between his need and the freedom of action of others. There is no justification for monopoly other than that it makes information a scarce good instead of a free one, and that it enables high risks to be taken, by promising high rewards in the event of success. On its own, of course, no monopoly can produce any reward at all unless it is protecting something which is wanted by its appropriate public.

DIFFERENT RISKS

Since risks vary, the corresponding rewards must vary, and any monopoly which underwrites these awards should therefore vary. Classical Patents do not, and cannot meet this requirement. In broad terms, all the Patent systems of the world operate on minor variations of the two terms of apprenticeship of the earliest grants. These would amount to fourteen years (which was in fact the term prescribed in the British 1883 Act). The U.S. grant is for seventeen years, and there is a widespread tendency to move to a twenty-year term, as in the case of the European Patent. These terms, as they are administered, make no attempt to achieve the objective of matching monopoly granted to risk undertaken. In the U.S. case a Patent may be surrendered at any time; in countries where renewal fees are payable (and eventually in the European Patent also) the Patent will automatically lapse in the event of non-payment.

Such provisions for varying the effective term, cannot be claimed to be in any sense for the purpose of the "matching" which is desirable. If the life of a Patent is terminated by the Patentee before its full term, it will only be because it has become irrelevant. The subject matter of its protection has either ceased to be bought (for example, because it has been rendered obsolete by new technology) or the Patentee has concluded that it never will be bought in any serious way before the Patent expires. What cannot be questioned is that as long as a Patent is helping a Patentee to make money, he will keep it in force. Considered as monopolies, therefore, all Patents ignore the differences between risks undertaken to produce what they profess to protect.

TAXATION AND PATENT TERM

They also ignore massive changes over time in factors which have a bearing on the return which is received by an investor in innovation. For example, it is claimed that seven or eight of the seventeen years of a U.S. Patent are now of no value to a pharmaceutical firm, because of the growth of certification procedures which did not exist when the Patent Act was passed in 1952. Far more important is that over a much longer time-scale, no attempt has been made to take account of the growth of taxation of profits and of the derived market power of Trade Unions, both of which greatly reduce the investor's residual return, which is the only thing that can justify his investment.

To be fair to Patent administrations, lengthening of Patent terms in line with these changes would not have had the desired effect, since a parallel development has been a general shortening of product life cycles. Even if it might now take a fifty-year Patent monopoly to produce the same theoretical after-tax return as a fourteen-year Patent a century ago actually did, this is quite irrelevant if the product protected will now be rendered obsolete in five years, either by competition, or by the Patentee's own ongoing R&D. There can be little doubt, therefore, that part of the relative decline in the use of Patents for underwriting innovation, is attributable to the maintenance of the monopoly protection in a substantially static form, over a long period during which several factors affecting return on investment have changed radically.

ADVANTAGES OF VARIABLE TERMS

The potential benefit, therefore, obtainable by making Patent and similar monopolies match their subject matter better than they do at present, must be correspondingly great. The Innovation Patent follows the logic of a variable term to its conclusion, by prescribing that terms should be explicitly related to the "innovative capacity" of the individual applicant. Similar capacity in a potential licensee is also a major consideration in setting licence terms for an Innovation Patent, in which the granting Authority is intimately involved.

These features offer particularly interesting arrangements for joint ventures between large and small firms, and between private inventors and firms of any size. The Innovation Warrant makes the term subordinate to public policing of incontestable grants, which, it is believed, would make short terms, compatible with modern product life cycles, possible. Both can claim to offer incomparably better matching of monopoly to risk undertaken, than the classical Patent system. In doing so, they cannot fail to improve the level of private returns to investment in innovation correspondingly.

The variable term faces up frankly to the need for a monopoly grant to correspond adequately to the difficulties which face an investor in innovation. The lack of direct protection for innovation has in fact led to the use of other methods of market power to fill the void created by the way in which Patents abandoned their traditional role.

This takes place in ways which were never intended when these other types were established. In Germany, application for a Petty Patent simultaneously with that for a "full" Patent is common, in an attempt to fill one of the gaps in Patent protection. The use of Registered Design protection, which was introduced with only aesthetic values in mind, for the protection of functional items such as engineering components, is a similar device to compensate for the lack of direct protection for innovation. So is the use of copyright protection for computer programs, as in the recent U.S. Appeal Court judgment in the Apple Computer case. Nothing could be further from the variable term of both the proposals of Chapters II and III, which is so carefully matched to the actual need for protection, than the application of a copyright protection term of a lifetime plus fifty years to something whose life-cycle is probably no more than one-tenth of this.

RELIEF OF UNEMPLOYMENT

Direct protection of innovation would contribute to a solution of the pressing problem of unemployment in three main ways:–

Firstly, by making innovation more profitable, it would generate an upsurge in investment of the only type that can lead to jobs that are viable and lasting. It is not necessary to be an uncritical devotee of "long cycle" economic theories, to be able to accept that every past recovery from depression has been associated with just such a burst of investment in new businesses and new industries.

As an example, in his book on long cycles, Rostow reported his estimates to show

"that authentic investment requirements in resource-related fields exceed the levels of investment required for sustained full employment over the next decade, and would require the setting of priorities". (7)

Clearly, however, the investments to obtain this desirable result will involve risk, and unless investors are convinced that they will be able to prevent enough of the returns from "leaking away" to competitors and others, they will not make them. "Authentic" investment in Rostow's sense, can only mean investment of which the bulk goes into innovation.

Secondly, Japanese productivity has been a significant factor in the growth of unemployment in other countries. The special contribution which direct protection of innovation could make to matching Japanese Capability market power, especially for incremental innovation, will be discussed below. Thirdly, direct protection of innovation would reduce the attraction of labour-saving as compared with output-increasing innovations. From a set of 1916 innovations reported by Franko, no less than 40% in the U.S. and 13% in Europe (only 6% in Japan) were labour-saving. (8) Such a weight of innovatory effort on actually *eliminating* jobs, must be a strong contributory factor in the widespread growth of unemployment in recent years.

DERIVED MARKET POWER

In order to understand what has happened, and how direct protection of innovation could help, it is necessary to add to the three types of market power outlined in Chapter I, a fourth type – *derived* market power. Like all similar power, this is the capacity to keep others out, and it relates particularly to the power of organized workers, not just to withdraw their own labour, but, by custom, by picket, or by threat of extending a dispute, to prevent others from taking their place. This power to keep others out is *derived*, since it depends upon the primary market power of the business. If that business fails, there is nothing for derived market power to act upon, and it is extinguished also.

Now, the stronger the primary market power, the more "rent" it can extract from the economy. There are then more resources available for derived market power to capture and divert into salaries and wages, instead of being paid to shareholders, retained in the business in order to develop it, or even invested in innovation. Wages and salaries therefore come to incorporate a "rent" element also. The inevitable result is to push the price of labour to a level which is above the price which would clear the market for it. It is not suggested that this is the only reason for unemployment, but it is certainly a contributory one. And the significant point for the present discussion, is that this reason *could not exist* without strong primary market power, since otherwise derived market power, however strong, could not push the price of labour much above the market-clearing level. It is the absence of a constant stream of new firms, founded by new men upon new ideas, that permits such a basically static pattern of industrial relations (which looks like conflict, but which is in reality a form of "adversary collusion") to exist. There can be strong primary power without strong derived power, but the converse is not the case. Strong power to keep other workers out of labour markets, depends upon the existence of strong power to keep other businesses out of product markets.

EFFECT OF CONCENTRATION

As explained in Chapter I, the line of evolution actually taken by the Patent system has made Patents a reinforcement of other types of market power instead of an alternative to them. This has undoubtedly contributed to the concentration of primary market power, since the more alternative arrangements for it there are, the more diffused market power must be. If primary market power is heavily concentrated, it becomes correspondingly more capable of extracting large rents from the economy. These rents become the target of derived market power, and to the extent that they are diverted into wages, workers are pricing their fellows out of jobs. Further, the obvious response on the part of those who control the primary market power, is then to invest in labour-saving innovations, and their market power enables them to do this successfully. Unemployment therefore has at least two roots in strong *primary* market power. One of these is that it enables wages and salaries to rise beyond the market-clearing level; the second is the incentive to replace people by machines, which high wages give to managements. Both roots are nourished from the way firms are more insulated than they should be, from the pressures that can only come from new entrants, or the threat of new entrants, to their markets.

As a means of encouraging such new entrants a Patent for invention is a poor basis on which to establish a new firm, or for an existing firm to break into a new field, because the protection it offers to innovation is so inadequate. But if the protection were to be given *directly* to innovation, this situation would be changed for the better. Immediately, it would be economically attractive to establish new firms to carry through innovations, instead of having growth take place only in established firms. These new firms would change the environment of the established ones radically, by correspondingly reducing their primary market power. Direct protection of innovation, indeed, appears to be necessary for the competitive process to work effectively. With it, Schumpeter's famous description of "creative destruction" has a vivid contemporary relevance, not merely an historical one. What counts is -

"the competition from the new commodity, the new technology, the new source of supply, the new type of organization, which strikes not at the margins of the profits and the outputs of the existing firms, but at their foundations and at their very lives. This kind of competition is as much more effective than the other as a bombardment is in comparison with forcing a door, and so much more important, that it becomes a matter of comparative indifference whether competition in the ordinary sense functions more or less promptly; the powerful lever that in the long run expands output and brings down prices is in any case made of other stuff". (9)

The absence of the "powerful lever" in recent decades has made static output and rising prices more typical of economic life, and this, in turn, can only reflect the absence of direct protection of innovation.

COMPETITION POLICY

The concentration of market power, of which the subordination of Patents to other forms of it, is a symptom, has obvious disadvantages. From the Sherman Act at the end of the nineteenth century in the United States onwards, therefore, Governments have been trying to reduce the harmful effects of this concentration. Their efforts have resulted in Anti-Trust legislation and various types of Competition Policy.

All such attempts to discipline market power have one characteristic in common: They apply only *after* the market power has grown to what is considered to be a socially unacceptable level. This means that Civil Servants are opposed to business men in a contest which the business men must always win. All the initiatives are taken by them, and the Civil Servants can only react; they are more highly motivated to gain the rewards of market power than Civil Servants are to control it; and their organizations will be less likely to be affected by bureaucratic inertia. As at present operated, Competition Policy ignores the reality that business men can only be disciplined by other business men, that is, by the actuality or the threat of new entrants to their markets.

There is therefore the possibility of a quite new kind of competition policy, which would operate *ex ante*, that is, by generating the conditions which will multiply such entrants, to strike at existing firms "at their foundations and at their very lives" as Schumpeter put it. In such a new policy, direct protection of innovation would be a vital factor, since it would provide the basis for the establishment of new firms, or for the expansion by existing firms into new areas. The failures of Competition Policy up to now, can be attributed to a significant extent, to the way in which protection of innovation was permitted to become only indirect. As mentioned above, this meant that growth tended to take place within existing firms rather than through the establishment of new ones. And it is the pressure from such new firms either actually or through the danger that they will come into being, that is the only force that can cause established firms to behave as the authors of Competition Policy want them to do.

MORE EMPLOYMENT

From the standpoint of employment, three results would follow from establishing direct protection of innovation:–
– Innovating firms are less efficient in their use of labour in their earlier stages than in their later ones. Only when their innovatory thrust is over, can they turn their attention fully to minimising labour and other costs. For a given output, therefore, they give more employment than established and mature production does.
– Reduced primary market power of established firms would make derived market power less able to push the price of labour beyond the market-clearing level. To the extent that this is a cause of unemployment, therefore, it would be weakened. It should be noted that this is true, even if the derived

market power remains as strong as before. Those who control it will be careful not to exercise it fully in the changed circumstances, lest they should reinforce the effect of external competition on the viability of the firm.
- Movement of wage rates towards the market-clearing level, reduces the incentive for those who control the primary market power to invest in labour-saving innovations, instead of ouput-increasing ones.

On all counts, therefore, it seems likely that direct protection of innovation could make a positive contribution to the world-wide problem of unemployment. It would do this primarily by changing the investment climate so that investors could expect to capture the rewards from their investment. As well as this, it would inhibit people from pricing themselves out of jobs, reduce the attraction of investment in labour-saving innovations, and direct investment towards bringing about higher output at lower prices.

EFFECT ON DIFFERENT SIZES OF FIRM

The shift from direct to indirect protection of innovation by the Patent system, has led to indulgence of the large firm, in spite of its disadvantages from several points of view. Originally, size was defended on the ground of scale economies; when empirical research undermined this, the ground of defence became the need for innovation, and the alleged impossibility of having it, except from firms with great resources. This is in fact largely true at present, but only because Patents protect innovation indirectly – and poorly. This means that the other types of market power have to carry the main burden of protection. It is of the nature of these other types that size is important to the protection they can give. A multi-plant firm can always use price to defeat a single-plant one.(10) The barrier to entry of an advertising appropriation may owe more to its sheer size than it does to any "quality" in the advertising. But size of firm need not enter into use of the principle of Patenting. Even the classical Patent monopoly is granted on the disclosure in the Specification, without any reference to what size of firm is going to turn it into reality, or even if this is ever to be done at all. A moment's reflection will make it clear, too, that if a Patent is performing its work of protection properly, the result should be altogether independent of firm size. The grant of a monopoly in exchange for a "teaching" (the ostensible intention of the classical Patent system as it has actually developed) should mean the same for a small local firm, as for a multi-national one. As discussed in Chapter III, however, it means nothing of the kind. Because small- and medium-sized firms have neither secure Patent protection, nor the other types of market power, they cannot rationally invest in innovation.

It is not only because such firms are more suited to some kinds of innovation than large firms are, for one thing because of their lack of bureaucracy, that this has serious consequences. It also deprives the large firms of stimuli to innovate. Since Patent protection on its own works so badly, large firms do not have to worry about being first into any particular field; they know that their entry will not be blocked, and that they will still obtain most of the market as a result of their capability and/or persuasive market power. It makes no sense at all for small- and medium-sized firms to get involved in innovation under such conditions, with the result that frequently the innovation is left for one of the large firm's (often foreign) rivals to undertake.

DIRECT PROTECTION AND FIRM SIZE

The situation would be altogether different with direct protection of innovation, according to either of the proposals described, or any amalgam of them. It is an interesting paradox that the classical Patent system theoretically ignores firm size, but ends up being completely subordinate to it, and contributing to it. Both proposals for direct protection, in contrast, take firm size into account from the outset, with the result that the protection they offer can stand on its own, and is no better or worse for a large firm than a small one. The Innovation Patent explicitly takes the "innovative capacity" of a Patentee or Licensee into consideration when settling the term of his monopoly or licence, and size would clearly be an important factor in this. For the Innovation Warrant, size would be taken into account when assessing firm-related risk. Even more important in its case, of course, would be relieving the Warrant-holder of the burden of protecting his monopoly, since this immediately makes differences in capacity to pursue litigation irrelevant.

By either approach, therefore, small- and medium-sized firms could play their full part in innovation for the first time. The monopoly granted by an Innovation Patent or Warrant would be a sound basis for establishing a new firm, or expanding an existing one, to carry through an innovation. Direct protection would generate new opportunities for large firms also, which would be stimulated to innovate in quite new ways. They could no longer hang back from a new area of activity, secure in the knowledge that if a smaller, more dynamic firm finds a valuable lode there, they can mine it too.

The Innovation Patent or Warrant would provide an insurmountable barrier to entry, in the interest of the original innovator. Even when the term which has given a small- or medium-sized innovating firm its prescribed reward (e.g. when it has satisfied 50% of the market under the Innovation

Patent system) has expired, large firms may still find that their entry is rendered difficult by the Capability market power which the smaller firm has been enabled to build up during its period of monopoly.With direct protection of innovation, therefore, large firms would have to be very much more innovative than they are now, with a Patent system which they have effectively "domesticated" in their own interest.

"MANAGEMENT "GET-OUTS""

Another stimulus to large firms would be that in contrast to the present situation, if they were slow in pursuing a particular innovation, they would run the risk that some of their more energetic executives would be able to find the backing to leave and attempt the innovation themselves. This is difficult under present arrangements, when the classical Patent held by such a firm, would effectively have a power of veto on all innovation, even if the firm is effectively sterilising the invention by its inertia. This is the *de jure* situation in the U.S., and the *de facto* one almost everywhere else, because token "working" (or even no "working" at all) is acceptable in the classical Patent system. It forms no part of either of the schemes for direct protection of innovation which have been proposed. For the Innovation Patent, the actual "innovation object" must even exist, and the "first commercial act" must have been performed, before protection is granted; for the Innovation Warrant, commitment to, and making of, the investment is the essential condition of obtaining and maintaining protection.

Direct protection therefore places a premium upon actually innovating, rather than on collecting titles to protection of abstract "teachings". Frustrated innovators in large firms would not therefore be prevented from looking elsewhere under the proposed arrangements. These, in any event, would be a vastly better basis on which to raise venture capital than a classical Patent could ever be. Once again, the only valid response which the large firm could make would be to be more innovative than before, and this socially valuable result would also be to the credit of direct protection of innovation.

SMALL- AND MEDIUM-SIZED FIRMS

The advantages of direct protection of innovation are especially topical because of the considerable stress which is being laid at present on developing the small firms' section of a number of economies. The German Small Firm Law is typical of such efforts.

The stimulus for these has come from unemployment, coupled with awareness (especially from a widely discussed empirical study from the Massachusetts Institute of Technology) that large firms are unlikely to contribute much towards its solution. One reason for this, of course, is that so much of the innovatory effort of such firms has been directed towards actually *saving* labour.

If nothing is done about introducing direct protection for innovation there is the greatest danger that the measures being adopted by Governments and their agencies will actually result in widespread distress, without contributing much towards solving the unemployment problem. In contrast to Japan (of which much more will have to be said below) the small firm in Western countries is extremely vulnerable. By definition, it neither possesses any significant degree of Capability market power nor the resources for Persuasive market power. Patents are positively a dangerous temptation for such a firm. If the likely cost of litigation is weighted by the probability of having to go to Court to defend the supposed "protection", no Accountant could possibly regard investment on the basis of a Patent as justified by any small firm. Yet the existence of the Patent, reinforcing emotional factors, can lure a firm's owners to just such investment and eventual disappointment, if not bankruptcy.

Empirical studies in Britain show how great the danger is of expecting too much from the small firms sector, unless the law which underlies industrial structure is changed. The chance of a new manufacturing business surviving and growing to 100 employees within ten years is no more than 0.75%. No less than 30% of all new businesses actually fail within five years.

Under present conditions, therefore, Governmental agencies that are encouraging small-firm development are like Generals sending their soldiers into an attack with instructions that they will obtain their rifle and ammunition from a dead colleague out of the line which preceded them. If direct protection of innovation were to be introduced, however, the position of the small firm could be transformed. Its lack of capability and/or Persuasive market power now becomes irrelevant. Because of the valuable monopoly conferred by its Innovation Patent or Warrant, it is now the equal of the largest firm as far as one particular innovation is concerned. The matching of the Patent term to the risk undertaken, by either of the two methods proposed, multiplies its chances of obtaining the finance it needs. Workers with special skills, who would otherwise only consider a large firm as a potential employer, can now be offered a tempting combination of security and profit prospects by a small one. It is not an exaggeration, then, to claim that not alone would direct protection of innovation bring about a much better balance in the economy between small- medium- and large-sized firms, but it is essential if present policies to encourage small business development in particular, are not to do more harm than good.

RESPONSE TO JAPAN

One aspect of direct protection of innovation which deserves special consideration at the present time, is a creative response to Japanese economic pressure. This pressure has been touched upon earlier as a contributory cause of unemployment in Western countries. It has also affected Western innovation adversely, however, and, what is worse, it may be expected to intensify. Four main proposals have been advanced for dealing with it:–
– Adoption by Western firms of Japanese methods of working.
– Matching Japanese exports with specialised, high quality products, especially in the consumer goods area.
– Protection by import controls, such as tariffs and non-tariff barriers.
– Doing nothing in the hope that improved living standards and more Government intervention in the economy will cause Japanese productivity to fall to Western levels.
None of these is remotely adequate to the challenge. The Western tradition is intrinsically individualistic and its successes have depended upon individual creativity. Japanese success has arisen from a quite unusual degree of social cohesion, of which the loyalty of the worker to his firm, the linking of pay to age rather than performance, the "Ringi" system which gives the highest importance to consensus as a basis for action, and the closeness of relationships between Government and Industry, are only the best-known illustrations.

PRODUCTION SUPREMACY

All of these factors are ideally matched to the productive aspect of modern industries, especially those which are intrinsically large-scale or catering for mass-markets. It has been pointed out that with its large population in a relatively limited geographical area, which is also characterized by an altogether unusual degree of religious and cultural homogeneity, Japan might have been made for the kind of business organization which depends upon the production line. In contrast, the diversity of the Western world reflects the individualism of its traditional values. It is inconceivable that the social factors which are responsible for Japanese productivity, could ever be translated to Western industry on a large scale without destroying most of the characteristics in his society which make life worth living for a Westerner.

RECIPROCAL TRADE

The second suggestion, of attempting to sell high-quality, specialised goods in Japan, is good as far as it goes, but it merely scratches the surface of the problem. There is no hope of being able to balance Japanese exports ranging from steel to word-processors, with fabrics from the great Italian designers and Chateau-bottled wine from France. Moreover, the list of Western industries which are under attack from Japanese productivity grows steadily longer. Pharmaceuticals did not figure on it previously, for two very good reasons. Firstly, as discussed in Chapter I, Patents for invention give valuable protection to innovation in chemical products, even though it is indirect, because the "teaching" of a chemical Patent Specification can only be innovated commercially in a single way. Secondly, the purchase of so many drugs is subject to Governmental influence, if not to actual public funding, that the Japanese feared covert official retaliation if they made serious inroads into the markets of the established companies. Markets where such influence does not exist, such as cameras and consumer electronics, therefore had priority. Now, however, with the Japanese research thrust accelerating rapidly, pharmaceutical markets too, may expect to be deeply penetrated by their new products.

There is no reason to think that the expansion of the Japanese export drive will be self-limiting in any sense. In fact, it would be prudent to assume that as they encounter restrictions in the markets where their share is now considered to be excessive, they will pay more attention to "lateral" expansion into areas which they have previously left alone. Western countries would therefore be operating from a shrinking base in any attempt to match Japanese pressure by following this second prescription.

COULD THEY NOT RUN OUT OF STEAM?

Hoping that Japan's advantage in productivity will be eroded by success, is only possible for those who know nothing of that country's past. Business is always low-grade Art, and innovation involves the same kind of energy, differently directed, that goes into artistic creation. Behind our own Industrial

Revolution stand the religious and aesthetic achievements of the high Middle Ages and the Renaissance. Similarly, Japanese technological achievements rest upon an outstandingly rich foundation in the Art and craftsmanship of the Tokugawa centuries – the period of isolation, before their decision to come to grips with the new knowledge of the West. The religious links with that past remain unbroken, and its Art is still reflected in the lives of the Japanese people; consequently, their expression in the contemporary idiom of technology is unlikely to lose its dynamism.

IMPORT CONTROLS

Protection by tariffs or by the more subtle devices used by the Japanese themselves, is often the most likely defence means to be chosen, but it is an acceptance of defeat. In itself, it does nothing to make the industry that is protected more innovative – in fact it ossifies it. Japanese products may be expected to continue to improve in quality, and to be produced at a lower price, compared with the protected ones, so that the latter will be finally forced out of all export markets. Indeed, the level of protection will have to be increased progressively, if even the home market is to be preserved.

IMPROVING INNOVATIVE POWER

Since the capacity to innovate is the key to being able to compete in export markets and to reduce import penetration at home, the proper response to any external economic pressure can only be to improve that capacity. Introducing direct protection of innovation would be particularly beneficial as a means of dealing with the Japanese case. This is because of the way in which the structure of Japanese industry is ideally suited to take advantage of the vacuum left by the practices which the Patent system actually developed. This point needs to be developed at some length.

ZAIBATSU-KEIRETSU

The Japanese firms whose names are known world-wide because of their products, are the Trading Companies and Bank-centred groups, especially the "big six" giant diversified firms like Mitsubishi, Mitsui and Sumitomo, known as Zaibatsu. Behind these are tens of thousands of small manufacturing firms, some of which are in a unique relationship with one of the large firms which trade overseas. Toyota, for example, has of the order of 30,000 component suppliers, by no means all of which would be related to it in this special way. This relationship is known as Keiretsu, and its paternalism corresponds to the lifetime employment in exchange for utter loyalty and dedication, which characertizes about a fifth of the total Japanese workforce, virtually all of it being in the very largest firms.

Where Keiretsu applies, the small firms "upstream" in the manufacturing process from their "patron" Zaibatsu, receive not just a constant stream of orders, but raw materials, credit, technical services and consultancy support as well. If markets change, the Zaibatsu firm will arrange for its Keiretsu client businesses to be provided with other components to manufacture, and they will also be carried over financially difficult periods. (11) What the Keiretsu firms provide in return is not just incomparably efficient sub-contracting, with minimal rejection rates for components and a degree of reliability which permits stocks to be reduced to hours' usage in some cases. *They also provide supremely effective arrangements for incremental innovation.*

INCREMENTAL INNOVATION

Until recently, the Japanese products which have relentlessly eaten into world market shares, have done so, not by radical innovations, but by countless small improvements which cumulatively bring about a final product which customers prefer to buy. Ease of servicing, lower use of energy, greatly increased reliability – all of these have been brought about by incremental innovations, with the increments in some cases being miniscule. This type of change is most likely to be achieved in the small firm, because it depends upon the closeness of management to the shop floor. Incremental innovations, whether in products or in the processes by which they are made, are not brought about by expansive strategic decisions in board-rooms or even in design offices. The possibility of them, and their realization, depends far more upon the skills, attitudes, knowledge and motivation of those who are actually making products, and it is the management of small firms which is in the closest possible contact with what is being made.

This is true of a small firm anywhere, but the Japanese small firm in a Keiretsu relationship has further unique advantages for incremental innovation over any Western counterpart. Its managers (frequently owners as well) are relieved of problems of survival, of finance, of raw material consistency and so on, and therefore can devote all their energies to making their individual products better. As a result of these Zaibatsu-Keiretsu arrangements, therefore, the ideal vehicle for incremental innovation is steered towards those innovations which are most acceptable in world markets. The products which incorporate the results are then promoted skilfully in those markets, backed by abundant resources. Zaibatsu-Keiretsu is, without doubt, the most effective means of bringing about economically relevant incremental change that the world has ever seen.

Nothing in the West remotely matches it, and in fact Western practice is in marked contrast. For example, in building up his great motor car business, now a part of British Leyland, William Morris maximised the use of sub-contractors. This was a deliberate policy of shifting the burdens of down-turns in a cyclical industry on to them. (12) In Western economies, sub-contractors will always be sacrificed in a Depression, and it is consequently not surprising that the components they supply do not incorporate anything like the same degree of incremental improvement that can be found in Japanese parts, and that "zero stock operation", which depends upon total supplier reliability, is merely fantasy for Western plant managers.

PATENT "VACUUM"

The relevance of all of this, of course, to the protection of innovation, is that the way in which the Patent system evolved, left a vacuum which might be said to have been made for the Japanese to fill with their mastery of incremental change. As discussed in Chapter I, the move from direct to indirect protection of innovation, resulted in Patents becoming "domesticated" in the interest of other types of market power, especially that of Capability. They virtually ceased to offer protection independently. Once "teaching" the novelty and not "performing" it became the basis for Patent protection, commercial arrangements for "filtering" ideas no longer operated, and Patent Offices had to devise some means of avoiding incurring ridicule by granting monopolies to worthless ideas, even if they were indeed new.

The method they adopted, the "inventive step" or the test of "non-obviousness", effectively removed incremental innovation from the realm of Patent protection. What characterizes this type of innovation more than anything else, is that once it has been done, nothing is easier than to reconstruct it from elements of prior art. Since by definition it is the type of innovation that emerges naturally and logically from what has gone before, it is totally vulnerable to the Patent Examiner's technique of "mosaicing" in this way. Adoption of the "inventive step criterion" to meet a problem created for

Patent administrations by granting Patents for invention instead of innovation, meant abandonment of Patent protection for much incremental innovation.

It may be objected that in fact the vast majority of patents granted are for incremental changes, not for path-breaking, radical innovations. This is correct, but the incremental changes which are patented, are not the ones which follow the "natural" line of evolution of a product in the direction which will cause it to be preferred by customers. Beggs analysed the Patenting activity of twenty industries in the U.S. from 1850 to 1940. His tentative conclusion was that the rate of change in Patenting was *inversely* proportional to the rate of change in value added.(13) As far as it goes, therefore, this supports the contention that the incremental innovations that are commercially worth-while, tend not to obtain their protection from the Patent system. Patents are more used by the fading, not the dynamic businesses.

NEMESIS OF "CAPABILITY MARKET POWER"

The inevitable result was that investment in this type of innovation could only be justified on the basis of some other type of protection. Capability market power was the most suitable, and it was readily available for the purpose, since the subordination of Patents to it had contributed to the growth of large firms, so that these came to dominate Western economies. From the point of view of such firms, this arrangement worked quite well before the Japanese thrust became strong, because of sophisticated market-sharing arrangements, to the efficient operation of which Patents contributed.

Once the full effect of Zaibatsu-Keiretsu on incremental innovation impinged on world markets, however, the result of leaving the protection of this type of innovation in the West to the market power of Capability only, was catastrophic. Now it was Japan which possessed Capability in the fullest measure, and consequently, it was Japan which now possessed the advantage in the only type of market power available to protect incremental innovation. Abandonment of its proper role by the Patent system, therefore, can be seen in retrospect to have made a present of this type of innovation to the Japanese. In comparison with this, even the other free gift, of the vast mass of technological information published through the world's Patent systems, which they monitored (and still monitor) to their great advantage, is of far less importance.

LOCATION OF INNOVATION

Reference was made earlier to empirical studies which show that where innovation will happen can be predicted if we can establish where the rewards of investing in it can be captured.(14) If a system of protection for innovation (e.g. a Patent system) is in being, but fails to develop so as to cover any particular area of activity, this does not mean that there will be no innovation in that area. It does mean that whatever innovation may take place there, will be what can be protected by alternative forms of market power. The corollary is that wherever there is market power in particular strength, it may be expected that the corresponding type of innovation will follow there. In the absence of Patent protection for incremental innovation, Capability market power becomes the protective means. Once Japan dominates in Capability, the focus of incremental innovation can then only be Japan.

PERSUASIVE MARKET POWER

In a comparable way, the United States possesses the market power of Persuasion to a unique extent, because of the unequalled scale of its firms' advertising appropriations and sales promotion expenditures. It is inevitable, therefore, that it should dominate innovation in products where this type of market power can give effective protection. What these products will be, will naturally depend upon the range of Patents. Where these give better or cheaper protection than Persuasive market power, they will naturally be used in preference to it. However, given the restricted protection level of modern Patents for invention, Persuasive market power may be expected to be used in the first instance for products in which the innovatory element is psychological. Here the physical element will have reached a plateau of development, resulting in standardization, so that innovation relates to means of favourably differentiating one brand from another.

The second type of innovation which can be expected to be protected by Persuasive market power (and consequently to be dominated by the U.S.) is that of products of any type which have a relatively short life cycle. In these, the protection consists of "secrecy plus lead time". The latter is the time it takes competitors to tool up for, manufacture, and get into distribution, an item which can be freely copied once it is on the market. Advertising and sales promotion do not change the length of the lead time; their function is to accelerate the rate of sales growth so that lead time is more valuable to the innovator, for example in getting an effective monopoly of retail shelf space.

Products with short life cycles are poorly protected by Patents as these are administered at present, so that Patents are rarely used for them. They are too difficult to obtain, too uncertain and too expensive to defend in the Courts if infringed. (The "inventive step" is a hard test for this type of product to pass, also).

Thirdly, there are innovations such as business systems, which Patent Offices have expressly refused to protect, even indirectly. Only those with Persuasive market power can afford to be concerned with these types of innovation. The firms with most of this type of power are to be found in the U.S. Under existing circumstances, therefore, that is where most of this type of innovation will be located.

RESTORING THE BALANCE

However, "existing circumstances" do not have to continue in force. As Chapters II and III have shown, direct protection for innovation is now a realistic possibility. For all Western countries, including the U.S., it would be a vastly better alternative to tariff or similar protection, as a means of coping with Japanese economic pressure. For European and other countries, in addition, it would open up possibilities of competing effectively with the U.S. in products with short life cycles or of the "business system" type.

Direct protection of innovation would grant monopolies that would be independent of all other kinds. Since it depends upon legislation, it is completely under the control of individual countries. It would make innovation possible without the need to call on any other type of market power. It would therefore immediately end the domination of certain types of innovation by particular countries because of the amount of other kinds of market power which businesses in those countries possess.

"CATCHING-UP"

In the case of Japan, for example, what is most needed now in many industries is not radical innovation, but "catching-up" on a large number of incremental improvements to existing products and in the way they are made.

This is where direct innovation protection could throw a lifeline to Western industry. All concerned see what the next step in evolution of a particular product will be, but they have also probably failed to make the last few incremental changes. Managements most likely felt (rightly) that if they tooled up for them, they would be submerged by the capability market power of a foreign (probably Japanese) firm before they could get any sort of return.

Assume, however, that they are now offered real, direct protection of the next incremental improvement, so that this fear is removed. A firm which otherwise faces annihilation in both home and export markets from the quality and price of competitors' products can now develop an investment plan, not only for improving on existing products, but also for manufacturing by the very latest techniques.

Because the improved product is not yet available on the market, the project would be eligible for protection, so the investment can be made in the secure knowledge that if technical success is achieved, then commensurate profits will follow. Every resource can then be devoted to achieving that technical success without the fear of sowing for others to reap. But since no customer will buy the further improved product if it does not also incorporate all earlier incremental changes, a comprehensive "catching-up" process must be undertaken and completed at the same time. This is not very difficult, since the foreign model is available to copy from.

In giving firms an opportunity to end up a step ahead of their previously invulnerable competitors, therefore, direct innovation protection also literally forces them to make up whatever ground they have previously lost. Moreover, it only protects them to the extent that their products are judged to be better than the foreign competition by the market, and it is strictly temporary. Compared with import controls, of whatever type, the protection given is of a creative kind, which, instead of ossifying industries, greatly helps them to make their products better.

INDIVIDUALISM, LAW AND CONTRACT

In the actual structure of Western industry, direct protection of innovation offers the best hope for arrangements which would be able to compete with Zaibatsu-Keiretsu. As discussed earlier, it would give small- and medium-sized firms an unprecedented degree of security in any innovative work they may undertake. With this as a basis, fruitful contractual arrangements could develop between them and firms with resources for international marketing. Strongly independent firms, linked for mutual advantage by legal means, are in line with the Western tradition of individualism, and could be expected to be correspondingly fruitful of successful innovation. In establishing a real alternative to Capability and Persuasive market power, any country or group of countries would be greatly reducing the disadvantage of their own industries in terms of innovation, relative to countries which possess those types of market power in abundance.

The most immediate effect of this would be to enable them to cope with Japanese competition. Even if direct protection of innovation cannot achieve this fully in respect of export markets, it could certainly do so in home markets, without having any of the harmful effects of import controls. This enormously beneficial result would come about if a combination which took what might be considered to be the best features of the proposals of Chapters II and III, was made available as a protection means. Although the protection would only cover the home market, the capability resulting from the new investment would also give an advantage in lead-time in export markets, and the concentration on technical success, also made possible by the terms of the grant, could only reinforce this advantage.

"ORIGINATIVE" INNOVATION

Economic pressure from Japan through superiority in incremental innovation is actual; pressure from Japanese "originative" innovation (the big "Schumpeterian" forward leaps) is looming. It has been shown how introducing direct protection of innovation could be the key to a powerful response to the former type of pressure. What could it do about the threat on the horizon?

In the early days of the Japanese export boom, it was frequently said "They are only able to copy". This is no longer heard, since they have shown how well they are able to *improve*, and indeed improve by orders of magnitude. It is now sometimes said (possibly by the same people who hope that the problem will be solved by the Japanese becoming as "soft" as Westerners through prosperity?) "Yes, but they cannot produce the inventions that require real originality". There is no reason whatever to think that this contemporary piece of wishful thinking is any more true than the earlier one was.

Even if Japan's social and educational structures may not seem apt to foster the individual genius who has been responsible for so much radical change in the Western past, other elements in those same structures provide exactly the support which "originative" innovation needs. The most important of these is a long time-scale for financial returns.

When Scherer and Ravenscraft studied the lag between R&D and the resulting profits in the U.S., they found that it tended to be between four and six years. This is probably typical of what can be achieved with current Western protection systems, because they rely so heavily on the Capability market power of the large firms, where management and ownership are separated, and whose shares are invariably traded on the Stock Exchange or Bourse. The resulting pressure for dividends makes positive decisions about projects with longer times to pay-off, difficult for managements of this type to take. The Pilkingtons have gone on record that Float Glass cost them so much before they got it right, that it could never have been successfully innovated if they had not been a family business. No outside shareholders would have stood for it.

LONG TIME-SCALE

In contrast, the Zaibatsu in Japan include a Bank or Banks in their closely-bound network, and of course, the comparably-sized Bank-centred Groups also involve similar close relationships. As a result, twenty-year finance for innovatory activity is not at all unusual. Nothing illustrates the longer time-scale to which the Japanese can work in innovation of the most radical kind than their steel industry. The efficiency of this, which has wreaked such devastation elsewhere in the world, owes a great deal to path-breaking fundamental research on the plasticity of materials by the Max-Planck Institute in Berlin. This was abandoned, on the ground that any practical application in the German steel industry was too far in the future to justify further work. It was taken over by the Japanese, and completed, with the practical results of which Essen, Sheffield and Pittsburgh are so painfully aware. The time-scale for innovation, which was too long for the German steel-makers, was within the range of their Japanese rivals.(15)

There is an analogy here with the incremental innovation problem. Zaibatsu-Keiretsu is a better means for achieving this than the predatory relationship between large and small firms in the Western world. Japanese industrial structures are also more suited to the long-term financing that originative innovation needs, than are the giant firms of the West whose managements have to respond to stockholders' requirements for quick results. It is easy to see, however, that the disadvantage of Western countries relative to Japan in each case, only exists in so far as the protection for either type of innovation depends upon the market power of Capability. This is the actual situation to a large extent, because of the way in which the Patent system developed so as to be only a reinforcement of Capability market power, instead of an alternative to it. It will be equally obvious that the introduction of direct protection for innovation could reduce the same disadvantages relative to Japan.

"VARIABLE TERM" ESSENTIAL

Because the question of time-scale is so important to successful originative innovation, one of the most valuable features of both proposals for direct protection of innovation (and which is quite fundamental in each of them) is the variable term of monopoly grant. Attention is re-directed to the earlier argument that such a variable term is logically impossible if protection is given to invention instead of innovation. If the classical Patent system were now to attempt to move in the direction of a variable term, it could only do so by having regard to characteristics of the firm innovating the invention, and abandoning its previous concentration on the "teaching" alone, to the exclusion of all questions of how (or even whether) it is to be innovated.

The variable term of either the Innovation Patent or the Innovation Warrant can be adjusted according to whatever time-scale is appropriate to innovations of different kinds. Partly because of this capacity to adjust, it offers protection of a kind that makes firm size largely irrelevant.

As a result, the innovation becomes much more capable of attracting finance on its own merits, even if the time to pay-off is seen to be a long one. The four- to six-year limitation in this respect, associated with the use of Capability market power, no longer determines what is, or is not, to be innovated. The virtually unlimited flexibility of the sophisticated variable term arrangements of the Innovation Patent, and features such as "Surrender and Re-grant" of the Innovation Warrant, are particularly appropriate for innovations with long time-scales.

For all these reasons, direct protection of innovation could not fail to have a marked effect on lengthening the average time to anticipated returns, of Western investment in innovation. To the extent that it did so, it would have made a contribution towards matching *future* pressure from radical Japanese innovation, that would be as important as what it can do to meet *actual* pressure from their capacity for incremental innovation.

TRANSFER OF TECHNOLOGY

Once "teaching" replaced "performing" the novelty as the Patentee's contribution to the exchange of the classical system, the implicit objective of that system could be nothing else from then on, except the spread of technological information.

What a Patentee gives in exchange for his monopoly, has always included a contribution to the diffusion of technology. When he had to "perform" the novelty, it was assumed that this "performance" would instruct others. When "teaching" replaced "performing", it was assumed that the necessary instruction would be included in the Patent disclosure, but the reality of the system does not at all match up to the second assumption. It is generally accepted that a disclosure which is capable of supporting a valid Patent grant, need not contain information (the "know-how') that is essential to putting the invention into practice commercially. This is so to the extent that the provision of such information is even regarded by some theorists as no longer the function of disclosure in the Patent system.

Reference was made in Chapter I to the fact that Canada paid four times as much for know-how as for Patent licences. Firms have an interest in depressing the value of a Patent and inflating that of "know-how", and where many licences are to subsidiary companies, this interest can have a particularly distorting effect on the figures reported. Even allowing for substantial bias, however, these Canadian figures confirm indications from other sources that

"the essential subject matter of licence agreements is almost invariably know-how...patents are normally a minor adjunct to agreements...seldom containing the essential information for working a new technique or process".

Perhaps the most telling piece of evidence available, however, is an authoritative submission to U.S. Senate Subcommittee hearings on a proposal to extend the Patent Law so as to enforce the provision of "know-how" in the Patent disclosure.

This argued that such a condition would effectively destroy the Patent system, since so many potential applicants would prefer to keep their information secret rather than look for Patent protection. In other words, if the "teaching" has to be complete enough to be effective, firms would not consider the Patent monopoly worth the exchange. So much for the classical Patent system as an effective means for the transfer of technology!

As in so many other ways, *direct* protection of innovation is much superior from this point of view. Under both proposals, disclosure must be complete. In the Innovation Patent, the actual innovation object is available for examination from the moment of grant, if not before. The provisions of the Innovation Warrant make protection certain, and since maintaining secrecy is uncertain, they could be expected to prevent attempts to maintain market power through secrecy. Both approaches insist upon the Disclosure being comprehensive, that is, it must include "know-how". But, because the monopolies they would offer would protect this "know-how", entrepreneurs would be willing to disclose it. Naturally, they will not do so for classical-type Patents, because "know-how" in a disclosure, being "obvious" or already known, is not protected. The result could only be that direct protection of innovation would contribute strongly to technological transfer in practical and effective ways.

REGIONAL IMBALANCES

The pattern of economic development in which growth is concentrated in the large company, results in other imbalances besides that between firms of different sizes. It also distorts the economic balance between regions within a country, as indeed it does between countries as a whole. Regional imbalance of this kind is inevitable as long as large size of firm is accepted as being necessary in order to bring about innovation. This in turn reflects acceptance that the only kinds of market power that are available to protect innovation, and effective in doing so, are those of Capability and Persuasion. Such acceptance is now widespread, and underlies the explicit policies which many Governments now have, for countering the resulting imbalance. In the EEC, Regional policy is also an attempt to compensate for the centripetal force of the common market.

The manner in which the imbalance is brought about, is most easily seen in the case of Persuasive market power. Where this is the primary means of protecting innovation, it means that the physical products have reached the stage of standardization. Their associated innovations are in the area of advertising and sales promotion, directed towards differentiating each product from its competitors in such a way that it will be preferred, and bought, even though it may be technically identical with them.

Like Banking, the administration of Persuasive market power gravitates to national and regional capital cities. That is where the media have their headquarters and where all the advertising agencies, public relations firms, marketing research organizations and specialist sales promotion Consultants are also located. It is not surprising, therefore, that wherever they may start, the managements of firms whose innovations are protected mainly by Persuasive market power, almost always end up running their businesses from the same cities.

Although the physical products are standardized, and so could theoretically be made anywhere, it is more convenient to conduct the necessary marketing/production liaison if the production facilities are in fairly close proximity. Also, since the products which rely most on Persuasive market power are mass-market, consumer-goods products, production is most economically located near the largest population centres. Both reasons direct economic growth towards such centres, especially if they are national or regional capitals, leaving the rest of the country correspondingly starved of it.

Where Capability market power is the dominant means of protection, similar forces are at work, even though they may be less visible. The centres where innovation takes place tend to be those with most amenities, the best educational facilities and proximity to other innovative activity – in short, the more pleasant parts to live in, of any particular country. Empirical work in Britain has shown quite clearly how strongly the location of actual innovations correlates with such factors.(16)

This might not matter so much if *production* of what had been innovated then tended to be located in the regions which are less attractive to mobile managers and researchers, but which have indigenous populations that are anxious to get the work. The same empirical studies, however, show that this is not the case. In respect of innovations emanating from firms which had more than one plant, the first commercial production had a very strong tendency to take place within a short range of the centralised decision-making and technical expertise of the group.

FINANCIAL TRANSFERS TO REGIONS

Such are the kinds of problem which Governments (and the EEC) have been trying to cope with by policies that typically rely upon financial incentives. These amount to current or capitalised disamenity subsidies to production. Although the sums involved are enormous, they have had nothing like the effect that might have been hoped for. It is recognised that much of the money spent is wasted. Where grants are given for buildings and/or plant, for example, the cost to the taxpayer is the same whether a project succeeds or fails, and failure is not infrequent.

In the light of the above analysis, the problem can be seen to be the pattern of *innovation* location, and the solution is equally obvious: Arrange for innovation to take place in the disadvantaged regions. On previous performance, production will then also be established close by, and the money generated will progressively close the "amenity gap" between one region and another. The difficulty about this solution is that as long as the only types of market power which are effective, tend to be concentrated in the areas of highest amenity, including national and regional capitals, that is where innovations will emerge. Patents cannot help, either, because they are now little more than a reinforcement of the market power of Capability.

DIRECT PROTECTION AGAIN

Yet once more, what could prove the answer to the problem is *direct* protection of innovation. It will have been noted that provision for regional differences is built into both of the practical proposals described in Chapters II and III. The Innovation Patent provides explicitly for the area of application to be smaller than a national territory. The ground-rules for administering its variable term of monopoly (which is established individually for each case) could presumably take cognisance of regional requirements. In the Innovation Warrant scheme, special "factors" are envisaged, which would be decided upon by Central Government, but applied by the Innovation Office to the term which emerges from its "project-related" and "firm-related" risk calculations, so as to give a longer term of monopoly in a particular region.

POSITIVE REGIONAL POLICY

Regional industrial policy based upon either or both of these proposals, could be highly sophisticated and capable of rapid adaptation to changing circumstances. The need to find alternative employment in a town where it is foreseen that a steel mill will have to close, for example, could result in exceptionally long monopoly terms being offered to protect innovatory activity there, and there only. Or Direct Patent/Warrant protection might extend to particular industries in certain areas, but not in others. None of these policies would require a penny in subsidy from either national or local Government; they would be virtually costless to administer; and the distortions and misdirection of energy that are inseparable from present regional incentive schemes would be eliminated. The industrial developments thus encouraged would be as organic and as soundly based as possible, because they would be undertaken and carried on throughout under the discipline of the market. Large differential advantages would indeed be offered to firms to invest in certain types of industry or in certain places, but actually realizing these advantages would depend upon the new products being bought in preference to those offered by competitors, so that profits are earned.

SCIENCE POLICY

It goes without saying that whatever direct protection of innovation could do to improve Regional policy, could also apply to Science policy. If the

Government decided that a particular discipline deserved more attention from innovators and entrepreneurs, it could issue instructions to the Innovation Office to award a longer term in appropriate cases. Some measure of the quality difference which exists between direct and indirect protection of innovation can be obtained by comparing such an incentive with the only known attempt made by the classical Patent system to differentiate its treatment of technologies. If an application before the U.S. Office could contribute to energy use or conservation, or to environment quality, it can be "advanced out of turn for the examination"!

INDUSTRIAL POLICY

Where Government is providing substantial finance for some new development, this is frequently allotted on a "sole source" basis on grounds of technical competence, specialisation, etc. Government support can therefore mean giving a single firm an effective monopoly of the commercial aspects of a project. Competitors are kept out of the development work by the deterrent that they will have to fund all the work they do from their own resources. They will again be kept out of the market, if the product turns out to be commercial, by the lead – time the Government-supported firm has gained. From many points of view, this is a makeshift surrogate for direct protection of innovation. It is inevitably bureaucratic, cautious and prolific of paperwork, and it has been shown to have a huge appetite for public funds to little economic effect.

Once again, comparison with the Japanese achievement shows that for commercial results, "it is corporately-funded R&D that counts".(17) Such corporate funding, in replacement of Government financial arrangements, could be obtained by the simple device of introducing direct protection of innovation. If business men can be assured of really worthwhile protection of their high-risk investments, so that in the event of technical success, they can capture the rewards of those investments, then they will have no need of Government funds.

UNDEVELOPED COUNTRIES

As was seen in Chapter I, indirect protection of innovation has been a cruel disappointment to the countries of the Third World, which were led to hope for so much from it. The experts in this field, Hiance and Plasseraud, have observed that a type of protection which is adapted to national technical development, and which is particularly in the interest of such countries, need not necessarily take the form of a classical Patent.(18) In fact, direct protection of innovation, according to either of the proposals advanced in Chapters II and III, or in a combination of some features of each, is intrinsically more adaptable to local conditions, irrespective of how advanced or primitive these may be. Novelty, according to both proposals, is no longer a question of what may have taken place on the other side of the world, in an environment that may be culturally and economically different to a far greater degree than even the geographical distance represents. Novelty is decided by what is, or is not, *locally* available.

There is no reason why a poor country should not even define novelty as meaning "not available in the ordinary course of trade *from indigenous production*". Such a definition would provide opportunities for native entrepreneurs to obtain vitally necessary experience in the discipline that is a pre-condition for manufacturing of even the simplest kind. These opportunities could only be availed of, however, to the extent that such entrepreneurs actually took the steps of production and marketing, with all their attendant risks, thus involving workers and others in the process of absorbing the same discipline. In contrast to the centralisation of economic activity which has actually widened the gap between poor and rich countries, Innovation Patents or Warrants would disperse it widely.

As an alternative to rewards from manipulating the State machinery, they would give them only for success in actual production, recognised as such by customers. In contrast to protectionism of the traditional kind, which has so often led precisely to lack of innovation, direct protection of innovation would be strictly temporary, with the term of monopoly carefully matched to the risks and the problems faced by the recipient. In the case of the Innovation Patent, they would even correspond to his "innovative capacity".

It is true that particular firms from advanced countries could suffer from lower export sales for their less sophisticated products, as a result – unless they themselves had taken the initiative about fulfilling the conditions of local production so as to obtain appropriate protection for this type of innovation. If. they preferred not to disperse their Capability market power in this way, then, of all the economic units in the world, such firms are best equipped to pay the resulting price. In any event, this could be quickly recouped by their own use of Direct Protection in their more developed markets. All in all, the new system responds to poor countries' need for trade rather than aid.

AN INSTRUMENT OF NATIONAL POLICY

At the root of all advocacy of indirect protection of innovation is the idea that what is given in exchange for the monopoly grant is the "teaching" of the novelty. So much is this the case, that the idea of absolute novelty is accepted, even by those to whom it is economically damaging, as necessary to provide adequate justification for giving a monopoly grant in exchange. Relative novelty appears to look for too little in return, from the Patentee. Absolute novelty could equally be called "universal" novelty, and, as such, it can never be the instrument of any economic policy other than a global one. Whether they realise it or not, therefore, countries which commit themselves to it, through the classical Patent system, are abdicating a degree of control over innovation within their territory. In contrast, adoption of direct protection of innovation, reasserts that control. Innovation Patents or Warrants are therefore instruments of national economic policy. As such, they would contribute strongly to the world-wide decentralisation of economic power that is essential if the gap between rich and poor nations is to be narrowed.

Further, since the "politicisation" of the World Intellectual Property Organization has resulted to a large extent from a widespread perception that the Paris Convention was not working to the benefit of the poor countries, so its "de-politicisation" depends upon a reversal of that perception. Such a reversal could come about through general adoption of direct protection of innovation. Then, use of the Convention by firms from the advanced countries in poorer ones, would demand investment from them for local manufacture, which could only be looked upon with favour. Arguments for leaving the Convention or for changing it radically would disappear, and W.I.P.O. could become a similar kind of organization to its predecessor, B.I.R.P.I.

CONCLUSION

The cumulative advantages of a system of direct protection of innovation are formidable. The principle of Patenting, it has been shown, is capable of exploitation in ways that go far beyond the classical Patent system. Except for chemicals, the protection which a Patent now gives to innovation is inadequate because it is indirect. The areas which have suffered most in the vacuum left by the actual evolution of the Patent system, are those of Mechanical and Electrical innovations. It is these areas, consequently, that would benefit most, and most rapidly, from the introduction of direct innovation protection.

In free-enterprise economies, the importance of Law is paramount, yet not nearly enough attention has been paid to getting the law right as it applies to innovation. All Government attempts to encourage innovation are attempts to make up for inadequacies in the law which relates to it. If we do not get the law right, these substitutes will not work; if we do get it right they are unnecessary.

And the historical precedents on this point are clear:–

Since the first industrial revolution, every recovery from economic depression has involved an identifiable quantum change in the law which regulates how business is done. It was not just railway-building that lifted economies out of the slump of the 1820s, it was, too, the great outburst of legislation for capital deployment; it was not just electricity, chemicals and the internal combustion engine which underwrote recovery from the even deeper depression at the end of the last century, but also the laws that brought the modern Patent system into being; before they became subordinated to Capability market power, Patents played a big part in getting the Science-based industries started. The Trade Marks Acts, on which mass-market advertising is based, share much of the credit for the growth of the consumer- and consumer-durable goods industries on which the long post-War period of prosperity was so largely based.

Reviving that prosperity now depends upon creative legislation to release another burst of business energy and investment. Direct protection of innovation is by far the strongest candidate for that legislation at the present time.

NOTES

1. von Hippel, E.: "Increasing Innovators' Returns from Innovation". Washington, D. C. (1981). Working Paper No. 1192-81 under National Science Foundation Grant No.PRA 80-119244.
2. Mansfield, E. et al.: "Social and Private Rates of Return from Industrial Innovations". Quarterly Journal of Economics 91 (1977) pp. 221–40.
3. Tewsbury, J. and others: "Measuring the Societal Benefits of Innovation". Science, 209 (1980) pp. 658–62. Nathan Associates Report to the National Science Foundation. Washington, D.C. 1978.
4. von Hippel op. cit. p.6.
5. Scherer, F.M.: R&D and Declining Productivity Growth. AEA Papers and Proceedings, May 1983, pp. 215–218.
6. London 1984 (H.M. Stationery Office).
7. Rostow, W.W.: Getting From Here to There. London. (1979) p. XXV.
8. Franko, L.: The Threat of the Japanese Multinationals. Chichester (1983) p.33.
9. Schumpeter, J.: Capitalism, Socialism and Democracy. London (1954 edn.) p. 84.
10. cf. Dewey, Donald: The Theory of Imperfect Competition - a Radical Reconstruction. New York 1969.
11. Yoshino, M.Y.: Japan's Multinational Enterprises. Cambridge, Mass. 1976.

12. Overy, W.J.: William Morris. Viscount Nuffield. London 1976.
13. Beggs, J.:"Long Run Trends in Patenting". Washington, D.C. 1982. National Bureau of Economic Research Working Paper No. 952.
14. von Hippel, E.: "Appropriability of Innovation Benefits as a Predictor of the Source of Innovation". Research Policy 11 (1982) pp. 95–115.
15. Gold, Bela. Productivity, Technology and Capital. Cambridge, Mass. (1979) p. 279.
16. Oakey, R.P. et al.: "The Spatial Distribution of Significant British Industrial Innovations". University of Newcastle upon Tyne Centre for Urban and Regional Development Studies (1979).
17. Franko, op. cit. p. 30.
18. Hiance, M. and Plasseraud, Y.: Brevets et Sous Développement. Paris (1972) p.93.

CHAPTER V

André Piatier

INNOVATION PATENT, INVENTION PATENT, OR BOTH?

Towards a Radical Reform of the Patent System

How simple life would be if all inventions could be covered by one unique and coherent definition! The same is true of innovations. (1)

The trouble with students of the law, with economists and even with technical people in this respect, stems from the trust placed in the vocabulary of current language, and from building rules and complicated projects on foundations which are ill-equipped to support them. Such artifacts are often constructed more on general ideas than sharply defined concepts – and these ideas can change their meaning from time to time.

There are two reasons for our difficulties with language in this context. Firstly, words wear out. Sooner or later their content of meaning becomes eroded. They can evolve towards abstraction or towards having a more concrete meaning. There are plenty of examples: Even in the 18th century, to be astonished was still "to be hit by a thunderbolt". For the Romans, scruples were pebbles in their sandals. Secondly, the things which the words describe are subject to change, but the words remain the same.

The terms "invention" and "innovation" have not escaped this evolutionary law. Nothing has changed more in a single century than these two concepts. We are hardly aware of it, however, because their "packaging"- the language we use concerning them – has remained the same. This phenomenon is observable in respect of most of the categories created by the human mind. Statistical nomenclature, for example, changes only slowly, so that a new phenomenon is only recognised in the real world when there has been "re-cataloguing".

Moreover, invention and innovation are often confused. Some recent studies find it very convenient that "innovations" may be covered by the invention patent. Facts fight a losing battle against the tyranny of categories generated by human mental activity. Over a long period, the word "patent" itself, has communicated very different meanings in different countries, and continues to do so today. W. Kingston has rightly pointed out above, that it originally meant the protection of an entrepreneur (called at the time a manufacturer) who brought into the country some industrial technique which had already succeeded somewhere else. His welcome from a particular State consisted of protection for establishment of the technique there. From the Renaissance onwards, first France and then England benefitted in this way from the inventive genius of the Italians and "indigenised" new ways of doing things in their respective countries. We can ask ourselves then, whether the current propositions on the subject of an innovation patent are not a return to the earlier way of looking at things.

However, it is worth noting that the former arrangements covered only a section of innovations, those which concerned the manufacture of objects already available in other places (they were innovations only in the geographical sense). To-day, economists and technologists favour "absolute" innovation, that is, the innovation of the most advanced countries. The old attitude was that of the less-developed countries of the time. It was also that of the United States, as is borne out by the declaration of George Washington to Congress on the 3rd January 1790. Legislation with regard to patents has as its aim

"to encourage efficiently the introduction of new and useful inventions from other countries, as well as the necessary knowledge and competence for making them an economic reality in this country." (Quoted from reference 1).

More than a century after George Washington's declaration, a high-ranking Japanese official expressed himself in his turn. By then, in other places the conception of the patent had changed, and now was aimed at the protection of the inventor, to give him the time to develop his invention and to extract new manufactures from it. The comment of this Japanese official was:

"We have looked around to see which are the greatest nations, so that we, too, may become a great nation. We have carefully studied, and have questioned, why is the U.S. such a great nation? We have made enquiries – and we have discovered that the reason was the existence of patents! Therefore, we too will have patents."(2)

Copying was the means they had adopted simply to overtake other countries, but for real innovation, they looked for a special stimulus.

In the industrialised countries, the patent – which is now essentially a patent for invention – has become decrepit and it is beyond doubt that it is now expedient and urgent to reform it. In the preceding chapters, W. Kingston and H. Kronz (I hope that I will be forgiven for ignoring the differences between their two contributions for the moment) propose two solutions, which, broadly speaking, can be characterised as follows:–

– the Innovation Warrant, designed to protect the person who invests with a view to launching a new product.

– the Innovation Patent, intended to give direct protection to a new product, which is ready to leave the factory.

These are the two propositions which I have to examine in this Chapter.

To begin with, it seems to me to be absolutely necessary to separate the two meanings of the word "innovation". This is at one and the same time *the new product*, that is, the result of exploiting an invention, and the *progression* by which the transition is made from invention to innovation, or otherwise the long process of development which separates them.

1. THE PROCESS OF INNOVATION

The analysis of this process allows us to evaluate the classical patent system. Formerly, when he first appeared on the scene, the inventor was often also the innovator. But since then, the two activities have been separated (1.1). Between the two, a development has taken place, which, little by little, has been extended. (1.2) To-day, this process is a succession of risks and costs. (1.3) Apart from this, we can take for granted, if not the disappearance of the independent inventor, at least his frequent integration, either into productive enterprises or into technical services or organizations. (1.4) These constitute a huge third source from which innovation flows. Such structural changes explain and justify research, either for new points of application of the classical patent, or for the juxtaposition of several systems of patents.

1.1. *The Inventor and the Innovator*

When it took the form which we still recognise at the present time, the patent applied, naturally enough, to the most important "agent", the inventor. It was necessary to protect him, since it was he who had found the idea thanks to which new products were going to appear on the market and the whole economy would find itself stimulated. In the nineteenth century, invention and innovation could still be considered as two closely-related operations. Most often, the inventor was an individual, and it was most often also he who was going to exploit his own invention. In numerous cases, he became an entrepreneur to manufacture the new product which he had discovered. Industrial history is full of these pioneers who started as craftsmen, but who then transformed themselves into small entrepreneurs – and who, sometimes, with luck, became the founders of great firms.

The beginnings of the motor car and of aviation illustrate this, with the pioneers working at the back of a garage or a workshop, to turn their "idea" into commercial reality. Some legendary figures are even more ancient. One example is Bernard Palissy (1499-1589) who, according to the Quillet dictionary, was a ceramic artist, a writer and a scholar. We are told that he sought to discover the secret of the Italian ceramic artists, and his story is a good illustration of the transition between the implantation of foreign technologies (covered by the old-style patent) and the independent discovery covered by the newer "invention" patent.

We are told that Palissy,

"poor but energetic and stubborn, sacrificed everything for his researches, going as far as burning his furniture and even the floorboards of his house. After more than fifteen years of trying, he succeeded in creating magnificent enamelled pottery. For his discovery, he was awarded the patent of invention for the King's rustic figurines."

Another pioneer, Denis Papin (1647-1714) discovered the possible uses of steam while observing a cooking utensil and built a steam engine. As a Protestant, he was forced to emigrate when the Edict of Nantes was revoked. Once he got on his feet again in Germany, he built a steamboat equipped with wheels. This "innovation" was destroyed by the boatmen of Fulda.

These two accounts teach some common lessons:

– The word "innovation" did not yet exist. The invention was the decisive fact. It preceded the appearance of the new product by only a short time.

– The inventor himself manufactured the product or the new machine which had been born in his imagination.

– Already, the fear of losing their jobs pushed workers to destroy machines. Since then there have been numerous examples of these violent acts in England and in France (for example, the weaving loom of Jacquart (1752-1824) which made it possible to weave mechanically, following a pattern. This was initially destroyed by the workers, but was eventually accepted by almost the entire industry of Lyons – 16,000 looms were installed in a matter of a few years. At the time, nobody spoke of innovation. All eyes were on the inventor. But, quickly, after the first industrial revolution, the distance between the invention and the actuality which it made possible, increased.

1.2 *Extension of the Innovation Process*

The simplest representation to-day of the succession of actions which border on innovation is a linear one: From Science to Technology, to

Industry. But, in fact, like the famous procession of the small town of Echternach in Luxembourg, (two steps forward, one step back) each stage can demand a return to the previous stages. For example, scientific and technical information can render an idea for a new product possible. Making progress with it, however, may demand a return to other scientific and technical information while at the same time calling for quite new technology, which itself can require its own tracking down of scientific and technical data.

The time is long past when the practical invention generally came before its scientific explanation (for example, steam power came a century before the beginnings of thermodynamics). To-day, scientific knowledge nearly always precedes applied research, which in its turn precedes the process of innovation. This process can be described in shorthand as a series of acts – research, development, laboratory prototype, tests, proof of the concept, factory proto-type, construction of production facilities, manufacture and commercialisa-tion. In each phase, fuller and more complicated networks of relationships are established between partners, since an individual person (the "inventor" of earlier times) can no longer follow the whole series of operations from one end to the other.

A study which is by no means the most up-to-date (U.S. Domestic Policy Review, 2 November 1979) introduces the problems and possible solutions of a process of innovation which is more or less equivalent (see diagram on next page). This calls attention to the possibility of weaknesses in the patent system and envisages a reform of this system. But in reality Patents are a problem not only at the phase this study entitled "Access to scientific and technical information", but also at the Research and Development stage, at the installa-tion of the productive facilities, etc. The diagram sums up the finding of this study which brings out clearly, once more, the weakening or even the disap-pearance of invention as such.

On the right of the diagram, everything relates to innovatory enterprise; on the left of it, to scientific and technical information, But, at each stage of the process there is need, in varying proportions and in different forms, for ideas, experience, information, skill and tools.

Patents can be concerned with one or other of these elements; they can correspond to various levels of progress of the innovation; they can be for a new product, "ready for the road", so to speak, or for an idea in its earliest form, requiring work to turn it into an innovation.

However, that is not the only distinction which may be useful to a consideration of patents.

–Innovation rarely appears by itself.

–Innovation is often a combination of innovations.

130

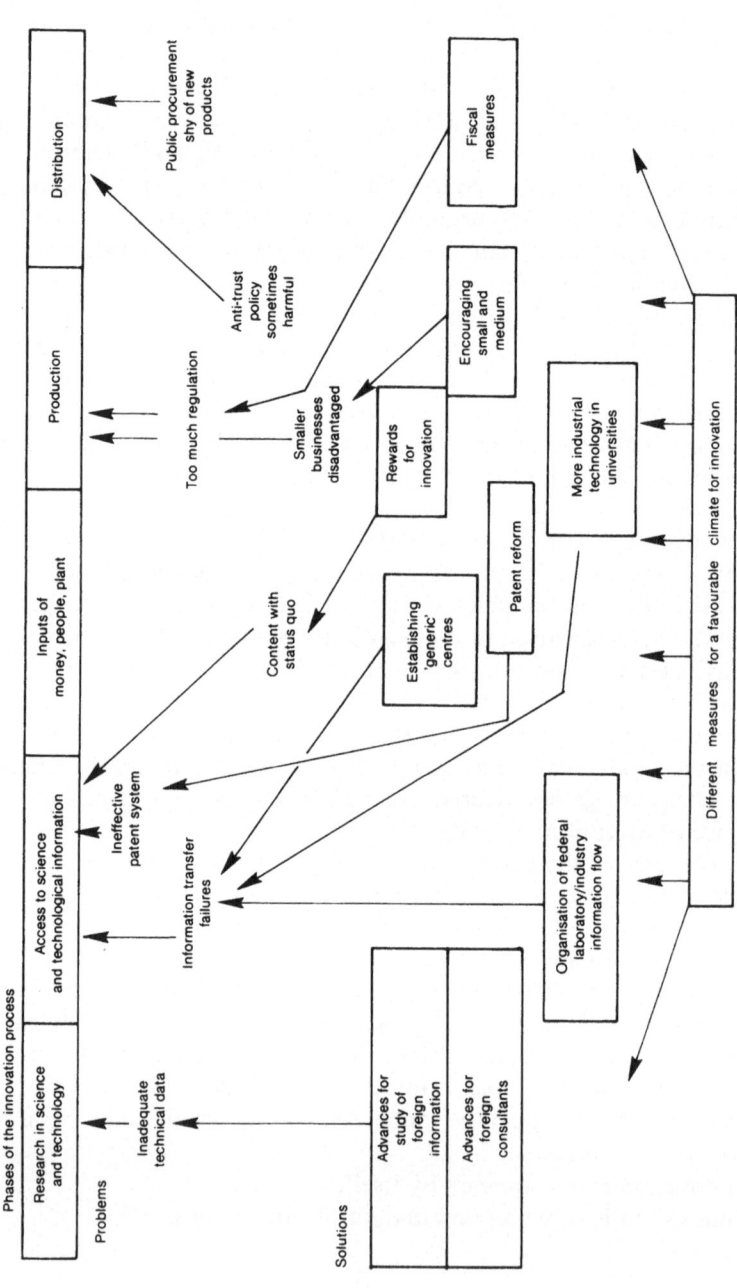

U S domestic policy review (2 November 1979)

Phases of the innovation process

| Research in science and technology | Access to science and technological information | Inputs of money, people, plant | Production | Distribution |

Problems

- Inadequate technical data
- Information transfer failures
- Ineffective patent system
- Content with status quo
- Smaller businesses disadvantaged
- Too much regulation
- Anti-trust policy sometimes harmful
- Public procurement shy of new products

Solutions

- Advances for study of foreign information
- Advances for foreign consultants
- Organisation of federal laboratory/industry information flow
- Establishing 'generic' centres
- Patent reform
- More industrial technology in universities
- Rewards for innovation
- Encouraging small and medium
- Fiscal measures

Different measures for a favourable climate for innovation

The first case is that of *adjacent innovations,* that is, innovations that are close to each other in "technological space". The concrete realization of a particular invention will entail during the different phases of the firm's inputs or of the development of the products, new demands on Research and Development for solutions to various problems. These might be new tooling or even new techniques for series production, new materials or new methods of extraction or procurement, the absolute necessity of lowering prices, or labour or capital cost, organization systems, etc. All of this work will result in "feed-back" from "downstream" in the direction of the "source".

The very high price of many new products at the time of their introduction, made it look as if they could never have much commercial success. The first ingots of Aluminium to be produced, for example, were more expensive than Platinum. But inventions followed by innovations in production techniques swiftly rendered the new metal competitive. It is the same to-day for certain techniques of cultures of plankton outside the marine environment. Their high selling price limits new products to the cosmetic field (end-products for consumers at high prices) using antibiotic plankton ingredients in very meagre quantities. But future innovations in hydraulic and civil engineering will open up the animal feedstuffs market to plankton-based products.

Another example was given at the 4th International Congress on Immunology in Paris, on 18th July 1980. Professor Capron then estimated that a single gram of Schiscotome vaccine, to prevent the disease of Bilharzia, cost 500 million old francs. Such a price eliminated any possibility of a major preventative operation against this disease. It will be interesting to see how long it will take until it can be produced at a sufficiently low price for the world's health services to be able to use it on a massive scale.

The second case is that of an uninterrupted succession of patented inventions, all dealing with the same problem. The majority of analyses generally seem to take it for granted that innovation is an isolated phenomenon. The reality, in fact, looks to be altogether different. The appearance of something new (a product or a process) creates both within the firm and around it, first of all, competitors, then suppliers and potential suppliers, all desiring to profit as much as possible from the new possibilities which are offered. The reactions of the "early adopters" of the innovation are also a source of stimulation.

As has been seen, research directed towards adaptation to the market is carried on through the multiplication of prototypes or models, by incremental improvements and also by genuine innovations on this or that particular point. Often the innovation of products is the first to appear and it is followed by innovations of processes. The rate of innovation does not decrease with the aging of the product. Marketing studies of new products show that the first wave of innovation has little chance of getting the product out of the embryonic phase. A supplementary effort is often necessary, which expresses itself through a second wave of innovation to break into the market and to launch the growth phase.

This was perfectly illustrated by Bjorn L. Basberg (4) in terms of the history of the adoption of the hauling-up slip in whaling boats. He was able to study, on the one hand the diffusion of this equipment in the Norwegian factory-ships, and on the other, the patents taken out in Norway on this subject, both over the period 1900–1960. As the diagrams on the next page show, the patents issued in the years 1905–1910 had practically no effect on the modification of the fleet and it was the bundle of patents issued during the years 1923–28 that started off the diffusion of the new equipment. By around 1935–36, 100% of the Norwegian fleet had been equipped with it.

The third case is that of a succession of innovations in the same undertaking. A. Tessier du Cros (5) describes the life of a firm as a succession of projects. New products, after all, grow old, and after a period of expansion, their turnover begins to decline. To survive this, the enterprise must launch innovations. Its growth is assured when the cumulative receipts from successive projects, result in an ascending curve. The patents which accompany these projects can differ greatly among themselves.

Since completing a study on obstacles to innovation (3) I have been wondering whether, in the same way as Friedrich List argued in the nineteenth century in favour of a system of protection for infant industries, it might not now be opportune to consider some kind of protective system for *infant innovations*? This would be of a different type – less linked to external trade, and more concerned with preventing attacks on organisms while they are still vulnerable, endeavouring to bring some balance into the supply of the necessary inputs, etc.

It is in this area that the innovation patent could find its justification.

1.3 *The Innovation Process: Additional Questions*

Yet a further diagram (3) aims at representing the process of innovation in a very simplified form, and at pinpointing different types of problems which arise, for those who are involved.

The phases of the process are represented by the rectangles, while the problems to be dealt with in each phase are in the ovals.

The solid line arrows show the successive phases of the transactions, while those with dotted lines show the problems posed.

It may be helpful for the continuation of this analysis, to observe here that

The relationship between patents and adoption of an innovation.

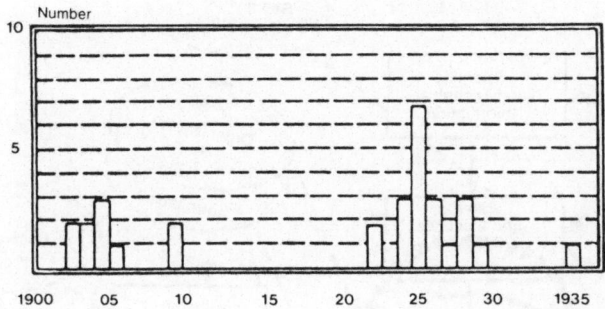

Patents granted in Norway in whaling concerning hauling-up devices.

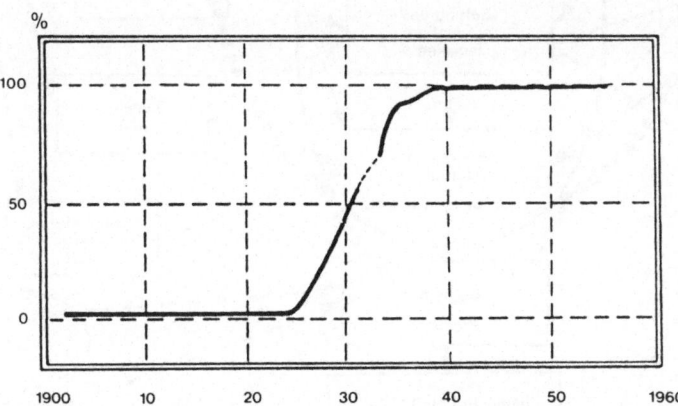

The adoption of the hauling-up slip. Percentage of the Norwegian floating factory ships in the Antarctic, 5 year moving average.

Source: The Whalingfleet Register, *The Norwegian Whaling Gazette*, and *The History of Modern Whaling*. From Bjørn Basberg: Patents, innovations and technological development in Norwegian Whaling. OCDE 15 september 1980, Science and Technology Indicators Conference.

134

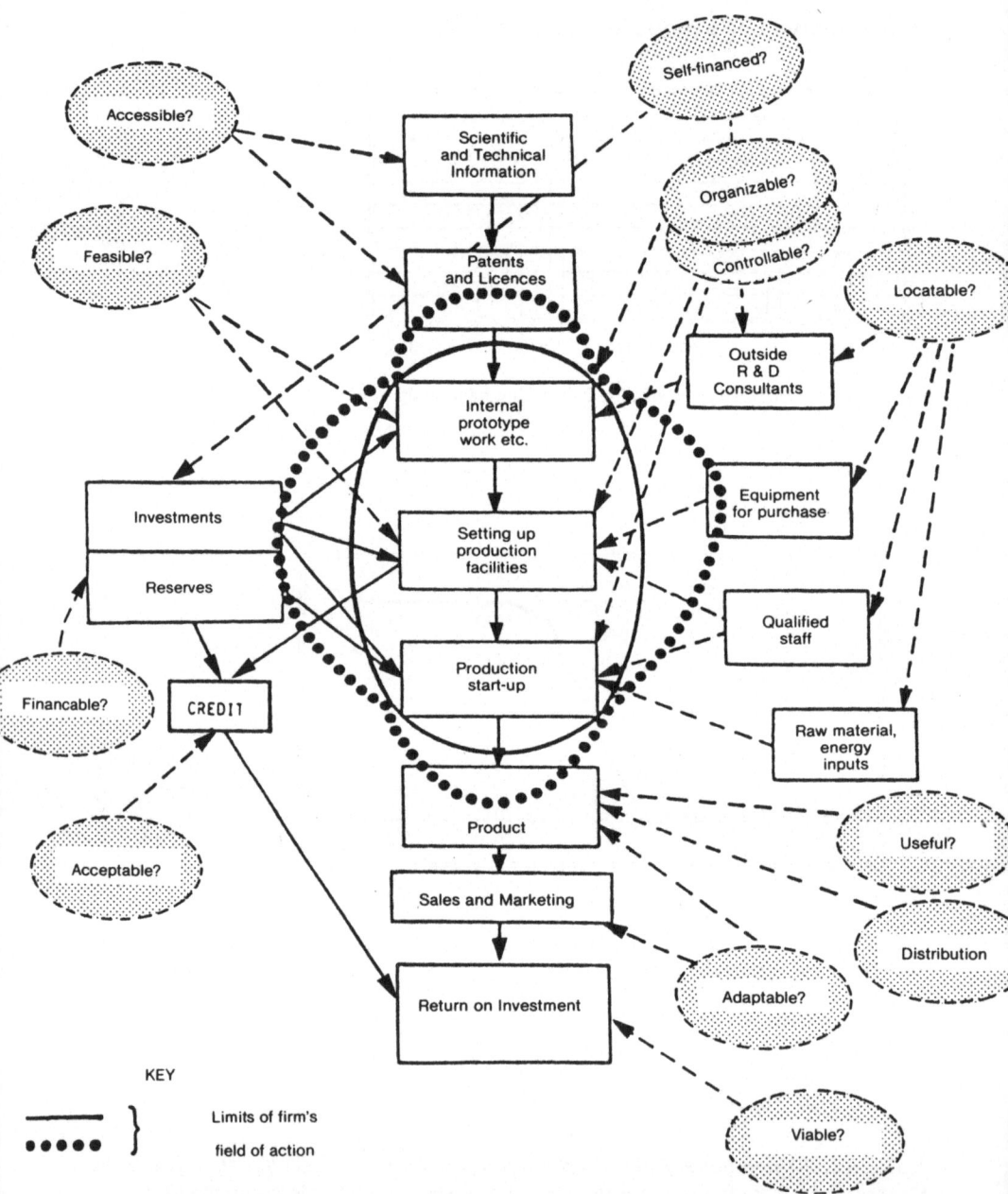

Questions raised by innovation. At each point, problems from Third Party actions and at the interfaces, are added to the firm's own problems.

these interested economic parties might be:

– a *single individual* if the inventor is an entrepreneur, or becomes one – or by extension, if the enterprise has itself actualised the invention which is the subject-matter of a patent.

– *two parties*, if at the beginning of the process there is an inventor, and at the end an entrepreneur.

– *several* parties, if between the inventor and the entrepreneur there has been a contribution from specialised agents of innovation (laboratories, research centres, consulting engineers etc.) It often happens that the patent which is lodged as a result of collective work is the property of a number of people or organizations.

In regular, day-to-day production, the problems do not arise (or if they do, it is only to a small degree). In contrast, for an innovation, the answers one wants to have are never evident. A single unfavourable reply to the first eleven questions can bring the process to a halt before we reach the final question: Is it a money-making proposition? Even if the most serious case is that of aborted innovation, the other possible difficulties – such as delays and increased costs – are by no means negligible. Patents (under their present form of patents for invention) are at the top of the diagram and investigation of how readily they can be obtained throws up several categories of problems:

– If the new idea comes into being during the innovation process, it is the conditions of the grant of the patent which must be examined, technically, financially, with what delay and with what guarantees of protection etc.

– If the innovation requires the acquisition of patents, the questions are different: Information on the existing patents which are at one's disposal, the validity of these patents, the work to be done in order to adapt them to the project, negotiations to acquire them or to get a licence etc., must be evaluated.

A double conclusion may be drawn from this section of this chapter:

(a) This presentation (of how innovation comes about) tends to endorse the propositions of W. Kingston and H. Kronz. New patents can be imagined all the way along the rectangles representing the process – and especially at "installation of production machinery" (the innovation warrant) and at the stage of the "product in being" (the innovation patent). The first would assure protection for the actualising of productive investments and the second, protection for the manufacture and sale of the resulting products.

(b) A different problem must be studied at the level of Research and Development, whether it takes place within the enterprise or outside it. Are the expenses which it entails sufficiently protected? Does the present system of patents (for invention) play a big enough role? It is by no means clear that it does, for several reasons:

– Since the definitive establishment of patents in the 19th century, the cost of Research and Development has grown and now accounts for a considerable proportion of the total cost of innovation. In numerous industries the cost of development outweighs the cost of research. In the aeronautical industry, figures from the U.S.A. show that development represents 97% of all Research and Development costs.

– The classical (invention) patent leaves the entrepreneur uncertain as to what the cost of its practical application will be. Certain patents cover an idea which is still relatively abstract and requires heavy outlays for its definitive embodiment, full-scale realization, or commencement of series production.

– In most accounting systems, the costs of Research and Development are not considered as investments – and very often they exceed the amount of the latter in the traditional meaning of the term, which covers establishment of the production facilities. A recent enquiry demonstrated this to be the case even in enterprises which do not represent the most advanced technologies. In a more general way, it can be said that technological evolution has considerably increased the share of intangible elements in the process of innovation. These elements, however, will be measurable in future – they are increasingly involved in commercial exchange, and they have become a new field of economic activity. Can the present invention patent cover both the intangible and material aspects?

1.4 *The Elimination of the Inventor from the Arrangements for Innovation, or his Integration into them*

When I carried out my first inquiry into innovation in France in 1981 (1) two questionnaires were involved, one for the innovator (the enterprise) and the other for the inventor. While more than 600 replies were obtained from innovatory enterprises, the diffusion of the inventor's questionnaire was a "non-starter". It really seems as if there are no longer many "private" inventors except perhaps a small residual group of independent, fertile and creative minds, whose size is measured by the number of patents still obtained by individuals.

In the course of these enquiries, I discovered only four who would be worthy of being called private inventors of a high technical standard. Others may exist, but they are few. The majority of highly creative minds are found either in productive enterprises, managing their research activities, or in consulting or research firms. Innovation Engineering is becoming a genuine new profession, distinct from classical engineering. While classical engineering confines itself to installing already known equipment, innovation engineering must invent new products as well as the production resources required by an industrial client.

In the area of high technology, the inventor is rarely an isolated figure. When he is found it is likely to be as head of an enterprise for manufacturing his own innovation. He can also involve himself in a firm which expects to benefit from his creativity. More often, it seems, the inventor is simply another staff member of a company. It can happen that after an important invention, he leaves it, with or without the agreement of his employer, to set up his own business.

As well, invention is a collective activity, either within a single team, or involving a number of individuals between which there is exchange of ideas, knowledge and resources. From that also, a double conclusion can be reached.
– A patent can be obtained collectively (for example, in France the patent for proteins obtained from petroleum, is held jointly by CNRS (the Government research body) and British Petroleum Ltd. This is what one can call horizontal teamwork, since the partners are working on the same phase of the development.
– But there is also need to take into account "vertical" teams, which are working on the successive phases of innovation. So we can therefore ask whether instead of successive patents (for invention – for investment – for a new product) there might not be an evolving patent whose successive forms would register progress as it was made. For example, there could be B1(a patent for invention), B2(covering R&D), B3 (Investment for production) and B4 (Cost of market development). Perhaps there could also be B5 and B6 for new applications of the innovation in other areas (for example, lasers in the textile or leather industries, in medical care etc.)

1.5 Emergence of a Third "Source" of Innovation

It is in this vast region that the national and international offices of industrial property are found, as well as patent agents, licencing consultants and the like. Within a single century, the interface between invention and innovation has become an immense new territory of novelty whose role is to administer, adapt and transmit knowledge, to bring about the passage from the idea to the "thing" (the final new product or the innovation) to control the transactions within this new group of intermediate "goods". Some of these involve information only, some are made up of information together with its material embodiment, and some are concrete, characterised by having a low information content. Any reform of the patent system must take into account this new domain of economic activity. In the coming decades, its role will become supremely important. This new domain must be properly inserted into the evolution of business.

Until the 19th century, the major part of production (as of employment) was concerned with final products. The manufacture of tools was very limited. The first industrial revolutions with their new sources of energy and their machines, reduced the proportion of total work associated with the production of consumer goods, but correspondingly developed the activities concerning "production goods". Full employment necessitated moving part of the labour force "downstream" on the one hand, in the direction of commercial activities and administration, and "upstream" on the other hand, towards secondary industry or goods to be used in further production.

The Industrial Revolution of the end of the 20th century is bringing with it the conditions of an "industrial exodus" analogous to the "flight from the land" of the last century. A large section of the active population, forced out of material production by automation and robots, will move further "upstream," to instal itself in this new "information economy" which has hitherto been largely ignored by those economists and politicians who should have been most aware of it.

That area, moreover, contains a mechanism that is crucial to the process of innovation, and which has been neglected until now. A balance sheet of the present situation of the information market should allow the interested parties, researchers, manufacturers and consultants, to become aware of the imperfections of their actions and their inter-relations and to benefit from examples of possible improvements offered by different means of intervention. It should also make it easier for those who are responsible for shaping public policies to improve their plans for innovation.

The highest creativity and the most original minds are not enough on their own to bring about economic and social development. It is necessary also to ensure that there is proper communication between those who generate new ideas and those who actually put new goods and services on the market. We do not know who the agents of this communication are. But there are more and more of them, and the tasks they undertake are probably increasingly complex. A superficial observation indicates a proliferation of professions and talents, with intense competition between specialists starting at different points to grasp new opportunities and assume new functions. It also reveals action that is disorganised, unco-ordinated and unfinished – something which the very diversity of "products" which are in circulation, also helps to explain.

A better approach to these intangible goods, that is, ideas, knowledge, and information, according to their nature, possibilities of exploitation, origin and objectives, is now indispensable. These intangibles have to be marketed. Who will do it? What is the product? How will it be sold? Where will it come from, and to where will it go? There is a market (supply and demand) for better information. Is the "product" offered adapted to the need? On the demand side, are users well informed as to the possibilities available to them? The classical patent, as well as the new patents suggested in this report, each have their place in this huge section. The idea of a "push" and a "pull" in this movement of ideas is not new. A more in-depth analysis, perhaps borrowing the techniques of commercial market research to use in this area, should be able to cope with the task of distinguishing the sellers' market from the buyers' market, according to whether the impetus comes from "offer" by researchers or "commission" by industrialists.

Too little attention has hitherto been given to this intermediate phase of innovation, which is, however, a decisive factor in eventual success. Those responsible for it have only been identified partially, although at least we know more about them than the means which they use, and the optimum deployment of these means. Some of these means are data banks, scientific and technical publications, indexes, all kinds of advice bureaux, conferences, seminars, technical exhibitions and trade fairs, public or professional organisations, university/industry liaison etc. Such a list is certainly incomplete, and like the reality, poorly organised. Specialists in new techniques are involved in the market for information, for example data processing experts. But in the international cases which we have been able to examine, some are preoccupied with providing a tool in the form of a document (for patent examination) similar to, if more comprehensive than what existed previously, while others find in patents no more than a means of selling additional "processing" time.

Transmission facilities now allow numerous enquirers to have direct access to data banks, and the number of these is growing very rapidly. Previously the relevant documents, whether computerised or not, were used almost exclusively by those who had brought them into existence. In the United States the growth of the computerised information market has been at an annual rate of 24% between 1979 ($1170 milllions) and 1985 ($4280 millions).(7) It is difficult to find estimates and forecasts of this type for Europe – even more so a breakdown according to the discipline where the information comes from, and by the type of work involved.(8) But it seems likely that in the years to come the growth rate of computerised information will definitely be weaker in Europe than in the U.S.(9)

It is absolutely necessary to transcend the problems of information, considered in its simplest form. The diffusion and exploitation of knowledge involves newer means and diversified techniques of which an analysis of the patent system offers a perfect example. At the beginning, and associated with the national offices established in every country, all that mattered was the protection of inventors; the job of the specialists was to draw up the patents, in other words, to give form to the inventor's idea. They needed two disciplines, one technical, to understand the novelty, and another legal, to delimit it in terms of formal rules. It is curious, but nonetheless certain, that the problem of the exploitation of the novelty did not bother these specialists. This had to look after itself, and the code of conduct of the Patent Agents' profession as well as legislation, prevented them – and still does prevent them – from helping the inventor in the commercialisation of his patents. Thus their activities aimed more at the elaboration of an intangible product (look at their painstaking researches into "priority"!) than at its exploitation.

This is why, in several countries, "licencing consultants" appeared, whose job is to find clients for patents, to organise contacts between those who supply and those who demand new ideas, and to draw up contracts to link them. Such "go-betweens" have to have a very wide range of skills, legal, technical, financial, and commercial, in macro-economics (to grasp the future of the innovation) and in micro-economics (to weigh up the capacity and the potential of the two partners). Further, the negotiation of patent rights often involves dealing with "know-how" and with the equipment necessary for production of the new product. It is these same intermediaries who carry responsibility for arranging for the exploitation of knowledge. Such people can work for a very wide industrial market, or, in contrast, can concentrate on well-defined targets, according to their size and policy.

Patent reform, then, must go beyond the limits of what can be patented at present and consider new things in areas where the rules are different (Pharmacology) or even reversed (Biology). Thus defined, the field of analysis represents the breadth of the phase commercialisation/diffusion of information. It is necessary also to state its length precisely. Between the supply (research) and the demand (industry) we may be able to find:
– absence of any intermediate agent. This is the case of the direct "producer-consumer" relationship (for example, University/Industry).
– one intermediate agent, who will often be an industrial advisor, a designer, a specialist in patents, licences etc.
– several intermediate agents who can act in a chain, in successive communications from one to the other, or, alternatively, simultaneously supplying different services obtained from several sources and offered to several clients.
– in the final category, the offer of knowledge can also come from elsewhere, from the industry itself, or from individuals – the industry's personnel, private inventors, or inventors in other fields or sectors of activity.

Every time that a "commercial agent" intervenes in the flow of knowledge, it may be necessary to determine how far the "product" he receives is specific to a particular stage; this "product" may not be usable by the industrial firm which undertakes final production. In that case, the intermediate agent carries out a function of transformation and adaptation. This function, which exists in all trade in tangible goods, must also be considered in information transfer. It, too, includes the functions of stockholding and transport. As Joel de Rosnay says, "the bulk of published work in the world accumulates daily at an accelerated rhythm".(11) A store of knowledge is coming into being, comparable to an information "deposit". As in the case of a deposit of petroleum or minerals, it is necessary to have powerful means in order to exploit it and get the best possible value out of it.

There is need for people who are specially trained in dialogue with academics, who have access to finance in a form that is able to be used as a source. (10) In practice, the exploitation of this deposit is hardly ever done at the "resource" stage, which is primarily in the Universities. It starts at the level of the technical centres and sources of research and development, that is, as often in public organisations of applied research as in industry itself. But this is also the business of the intermediate agents, as is proved by the growing part played by advice bureaux in the clientele of data bank users. Research on research is therefore becoming an essential function, not only at the industry level but also at that of the intermediate agents in knowledge transfer.(12)

Globally, every reform of the patent system must allow identification of who transmits and begins to exploit knowledge, considered as a resource for industry. It must take account of whatever existing information flows there may be. It must also help to detect blockages in these flows, misdirections, poor adaptation to the needs of the potential user, and excessive costs of all the activities which are sandwiched between Research and Industry.

It is not enough to have good research and good industry in order to end up with successful innovations. It is also necessary to reckon with the efficiency and productivity of the transactions involving the intangible elements. The new patents have an essential role to play in this.

2. THE NEW PRODUCT AND THE PROBLEMS OF THE INNOVATION PATENT

To wish to apply the patent idea to a new product or process without deep analysis is hardly likely to be very effective.

2.1 *A wide range of products and processes*

If the world of innovation comprised only genuinely new ideas, its landscape would be pretty deserted. In aeronautics, for example, after the propellor aeroplane there would be only the jet... and then, Challenger 40 years later. Such outstanding new ideas are behind the major innovations. They transform economic life, and they have the power of generating strong growth. This is not an appropriate forum in which to discuss the explanation of why these innovations tend to "cluster" around identifiable points in time, rather than emerging at random intervals.(13) Improvement or perfecting of a product or a tool, without being considered "major" from the technical angle, has also had a big effect. The coming into universal use of the metal ploughshare, for example, had major economic consequences, only observable with a century's hindsight.

Generalising the conclusions drawn from observations, discussions and examination of survey responses, it can be said that there are four categories of innovation whose "visibility" is not immediate:

"Unobserved" innovations: Those which change the performance, the quality, the precision, the viability of a product or a tool. Two models of a motor car, separated by an interval of 10 years, reflect a totally different "product" under the same name.

Hidden innovations: An industrial installation, machines, actually operating in a factory, can generally be kept more secret than whatever they produce.

Forgotten innovations: Too often, only physically tangible new objects are considered as innovations. This neglects intangible innovations; in the debacle that is European steelmaking, only three enterprises are profitmaking in any substantial sense. They have the same equipment and similar personnel who have the same qualifications as the others. The difference has been solely in organisation.

Unrecorded innovations: They exist but they are not adverted to. Neither the customs officer nor the statistician recognises them individually; they classify them in lists that already exist. It is only when the new product is widely produced or exchanged that a special category is created to enable it to be distinguished from others. The trouble is that the innovation is no longer anything new by the time it obtains separate recognition in the statistics.

All these categories reflect the uncertainties which trouble the observer of innovation. They probably are not of major importance for a new system of an innovation patent. However, it can be asked whether the manufacturers concerned would be interested in patenting "invisible" advances either in their output, or in their productive equipment. It is worth asking the question because, in the past, a frequent reproach concerning the invention patent was the uncertainty about the propensity to patent according to industrial sectors, countries and enterprises.

A new patent system will not be considered satisfactory unless it succeeds in covering all (or nearly all) innovations – and in constructing some sort of hierarchy (the idea of a hierarchy seems essential). With the invention patent we have at present, interpreting statistics is difficult: "Patents" cannot be aggregated because "two patents" may be worth either more than double, less than double, or the same as "a single" patent.

2.2 *A Typology of Innovations*

A typology seems indispensable to classify innovations and thus to establish categories for a new patent. In the preceding works (1) and (3) I have attempted to produce such a typology in the form of a matrix. This can be done if the criteria of innovation (novelty) are set out horizontally, and the points of application of the innovation are listed in vertical columns.
The criteria of the innovation.
It is indispensable to distinguish several criteria:–
– *The criteria of the nature of the innovation*:–
– new product
– improvement or perfecting of an old product
– accessory added to a product
– accessory incorporated into a product
– adaption to a new market
– adaption to new functions
– *Criteria of the technological level*:–
– very advanced technology
– advanced technology
– "State of the Art" design
– "Technological level" does not apply.
– *Criterion of novelty*: Is it –
– in the world for the first time?
– in Europe for the first time?
– in the country for the first time (but already realised elsewhere)?
– *Criteria of origin and destination*:
– innovation originating in a particular sector of activity
– innovation originating in other sectors
– innovation directed towards a particular sector of activity
– innovation directed towards other sectors of activity.

Classification of innovation by application (in columns) is a means of identifying where the new thing can be applied. Such a product can be distinguished in terms of its inputs, its conditions of manufacture and its distribution.

– the "product" can be a consumer good, an intermediate good, productive plant, a process, a service etc.

– its "inputs" comprise the technology (know-how), research and development acquired from outside etc., equipment, raw materials, partially-manufactured products, the labour aspect (qualifications of workers, their conditions of work etc).

– its "manufacture" comprises Research and Development carried on within the enterprise, construction of the plant, the machines used, its organisation, operating techniques, etc.

– its "distribution" covers all forms of commercialisation or of diffusion – new markets or a response to new needs, new functions etc.

Use of such a typology permits innovations to be divided into several categories and ranked in a hierarchy; for the effective administration of an innovation patent, this is necessary if innovations are not to be a confused and unintelligible mass. The objective of this typology is to track movements across sectors which have been obscure until now. To illustrate, a recent enquiry described how the toy industry was approaching the industries of advanced technology, through its use of electronic elements.(6) It is probable that technological transfers of this type could be covered by the innovation patent.

3. HOW CAN THE PATENT SYSTEM BE REFORMED?

The foregoing analysis of the process of innovation has shown how the distance between invention and innovation has increased. It is now so long, that concentration on the invention only is unjustified. But the analysis has also shown the complexity of the new innovation resource which develops "upstream" of actual industrial production; and at the same time, the presentation of a typology of new products showed the complexity of intervention at the "result" phase of innovative actions.

3.1 *A military analogy*

Here the model is that of artillery practice. What is needed is a weapon, a target and a good definition of what one is trying to achieve.

The weapon, for us, is the patent. Is it the only possible weapon? Can we not also make use of copyright, Trade Marks, or the utility model (or Petty patent)?

And if the patent remains the principal weapon, how can it be changed, or its efficiency multiplied? Hitherto, the target was the invention. The propositions of H. Kronz and W Kingston now identify two supplementary targets. Kingston's Innovation Warrant aims at the middle of the innovation process; in contrast, it is the end of this process, when the new product finally exists, is being manufactured and is actually ready for sale, that would be the area of Kronz's innovation patent.

An important question is whether the new systems should replace or complement existing patents (for inventions)? Kronz is on the side of replacement of the invention patent by the innovation patent, while admitting a possibility of their co-existence. Magnanimously, however, he would not forcibly suppress the invention patent – he would let it fall into disuse. Kingston proposes adding the system of the innovation warrant to that of the invention patent; its function would be to fill a void which is left by the existing patent system. He thinks that protection of invention is always necessary, but that it needs to be complemented by a system placed further "downstream". One can accept, a *priori*, that if we are talking about covering the entire field of innovation, it would be preferable to use the three systems at once:–
– "upstream", the existing invention patent;
– "midstream", the innovation warrant;
– "downstream", the innovation patent.

Further studies should therefore examine whether this multiplication of weapons is too costly... and too bureaucratic. In anticipation of these, all that can be asked is that reform should be open to the creation of "rapid-firing weapons". Too many national offices of Industrial Property seem to act as if eternity is the background to their work and too many tribunals seem to be content to stock a museum with contested patent claims.

The following diagram relates the three weapons to their targets. It illustrates that there is a degree of imprecision in the "shooting." The invention patent, as we know, covers a vast area, from the relatively abstract idea which is nevertheless well described, to the laboratory prototype. The innovation warrant has a narrower spread of fire, because it aims at the enterprise just when it is about to make a productive investment. But it will have difficulty in adapting itself to the large number of possible cases. These could range from where an applicant has to make an investment comprising the totality of productive requirements, through situations where he may already have ap-

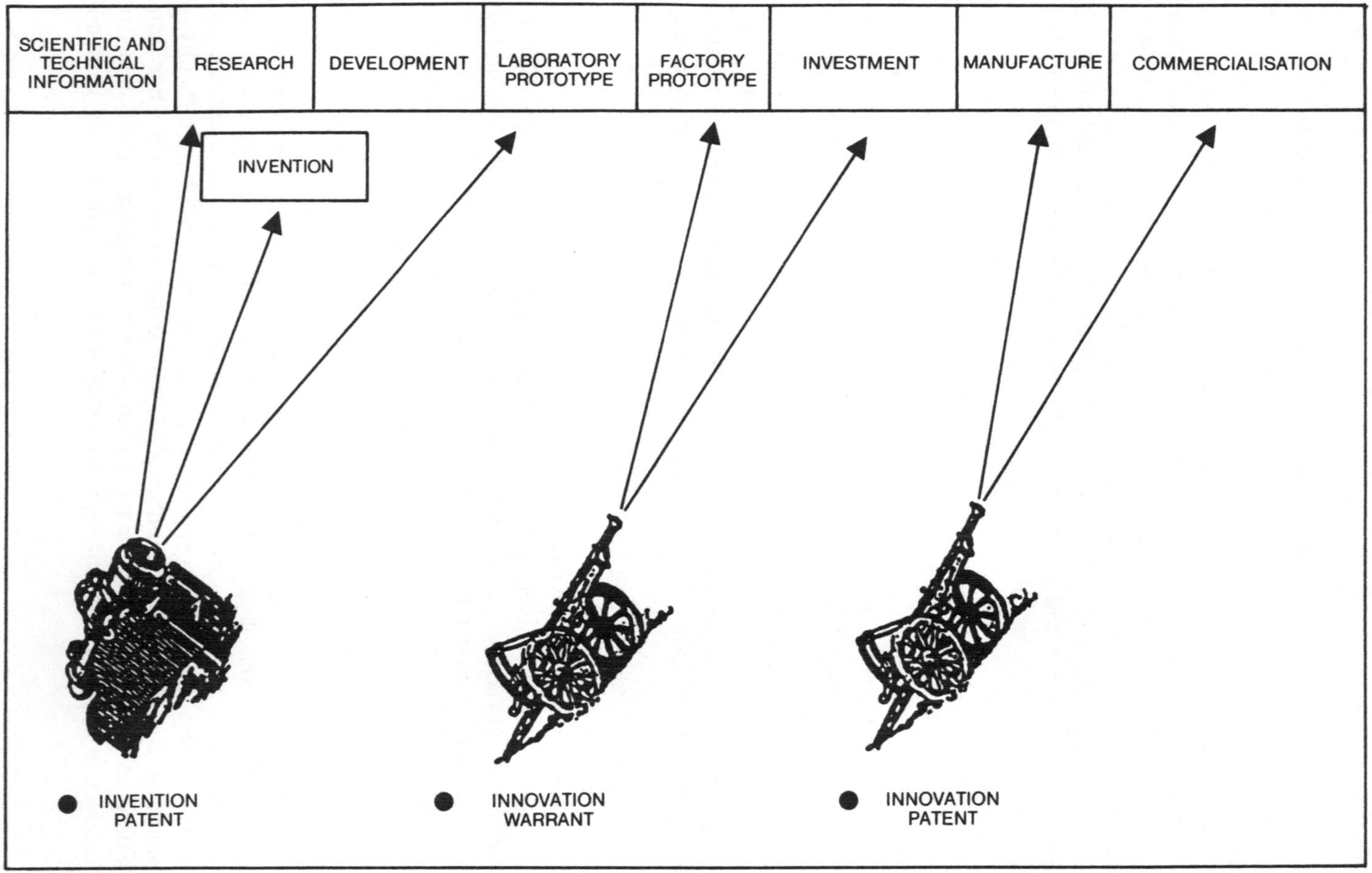
SCIENTIFIC AND TECHNICAL INFORMATION
RESEARCH
DEVELOPMENT
LABORATORY PROTOTYPE
FACTORY PROTOTYPE
INVESTMENT
MANUFACTURE
COMMERCIALISATION
INVENTION
INVENTION PATENT
INNOVATION WARRANT
INNOVATION PATENT

propriate facilities, or where such facilities are capable of adaptation, to cases where quite new plant would be needed, possibly in a new location, demanding the use of patented equipment of the most up-to-date type, etc.

The innovation patent, too, appears to have the drawback of covering a vast area, described in section 2.2 (the typology of innovations). How would it be possible to choose between what it is necessary to aim for and what should be left alone? A wide or a narrow focus will give very different targets.

The third aspect of the artillery analogy is that of expected results. Hitherto, the result which has been hoped for from the present patent system, is the protection of the inventor. Should nothing more be contemplated than extension of this protection to innovators? Such an objective seems to me to be both too grandiose and too limited. It goes too far because the inventor and the innovator are not beggars, nor are patents a social security system. It does not go far enough because what is at stake is not just the granting of a limited monopoly in a competitive work, but the simultaneous encouragement of human creativity and national economic growth.

3.2 *New objectives for a system offering diverse patents*

As the preceding diagram shows, problems of setting precise objectives arise in protecting innovation as well as invention, and in intervening when it is necessary to encourage investment. Here, I find myself in agreement with W. Kingston, but is protection of the successive agents of innovation the only valid objective? Like the author of the first four chapters of this book, I do not think so. Moreover, I have expressed the same opinion at the European Commission's meeting of experts on the 3rd and 4th November 1982 at Luxembourg, and also to the OECD. (14,15)

Over a century the possible subject-matter of a patent has changed radically. For a long time, the patent system had only to be a response to the problems of individuals. The first of these was the inventor; and the innovator will now be protected if the Kronz/Kingston suggestions are accepted by individual countries and by the appropriate international organisations. But more than protection of individuals is required now. To-day, innovation (and all that precedes it) is the business of States and even a matter of global economic development. Macroeconomic requirements are added to microeconomic needs (those of the individual inventor or the innovatory firm).

Since we know that industrial revolutions are generated by waves of innovations, we must now ask ourselves if the world economy will know how best to administer the potential of emerging innovations, electronic and information technology, telecommunications, rockets, satellites, biotechnology etc. which would enable us to leave behind the long crisis which began 10 years ago. (13) (16) A list (which cannot, of course, be exhaustive) of ambitions might include the following:–

– At the level of inventions:–

Stimulation of creativity and inventiveness. As A. Bouju (14) said, in protecting its holder, the patent no longer plays a stimulating role, either "upstream" (in terms of science and technology) or "downstream" (in terms of industry). It can be compared to the motorist's safety-belt; this may protect, but there is no incitement to grasp the steering-wheel and drive. Why could the invention patent not open the way to *exploitation* of the ideas to which it applies? Patent Offices and the private sector professionals should be encouraged to suggest, according to the inventor's wishes, either possible partners in the productive process (through systematic research leading to grant of licences) or means to enable him to go into production himself; the innovation warrant would be the natural successor of the first patent.

– In the course of the production process:–

To accelerate the passage from invention to innovation. The studies of G. Mensch, C. Marchetti and numerous other authors have shown that for radical innovations the intervening period can vary from 10 to 100 years. Recent enquiries show that for incremental innovations of the kind made so regularly as to be commonplace, the delay is often more than 2 years.

– At the end of the process:–

The innovation patent would confirm not only the technical originality of the new product, already manufactured and about to be commercialised, but also the existence of a market at the price determined by the seller.

At all stages the threefold patent system should:

– Improve the information available to inventors, investors and marketing people. A statistical reform of the patent system should provide a fundamental source for information. (17) (18) A compilation of resources available to be called on for research work, for adaptation and for development could be offered by the patent office to all those who have obtained a patent.

– Facilitate contracts and exchanges, either for the idea (invention) or for the manufactured product (innovation) or for "partial products" obtained during the course of interaction. For example, there might be a developed invention ready to be put into production because it has been proved, and its method of manufacture is already known. Or an invention which is already the subject of an innovation warrant might be considered as a "probable" innovation, worthy of investment to bring it to reality.

– Encourage financing of all the successive tasks included in the innovative process. The three patent systems, strategically located along the path between an idea and its realisation, should act as an Intelligence Service alerting the different categories of financiers on the importance, the soundness and the chances of success of the projects for which backing is required.

In the present state of relationships between bankers and innovators, I have attempted to distinguish between the different responses obtained when finance is sought. The most widespread opinion is that bankers do not, in fact, support innovation and that therefore it is necessary to find other means of finance, such as venture capital (which in my opinion has a very different role, more "upstream"' in the process). At the same time, many encounters between manufacturers and bankers, still leave the idea intact that the Western economy could not have grown at more than 6% per year since the end of the Second World War, unless the banks had been actively involved in financing innovation. It is easier to analyse the difficulties if these are looked at from the point of view of the market. This will identify what bankers will finance and what will require other means. For simplicity, let us imagine 7 categories of innovations:–

1st Category – The innovation is an improvement or refinement of an established product, widely in use. Since the market already exists, financing by a bank is virtually certain.

2nd Category – The innovation is a new product, intended to replace an established product, already widely used. Its price, convenience etc. give it an advantage over the old product. Again, bank finance poses no problem, although there could be some resistance if its mode of operation or its raw materials are unusual.

3rd Category – A production process, used for an established product, giving equal or better results. There is a good relationship between quality, performance, viability and precision of operation, and price. Bank financing is likely to be available, since this case falls into the area of the classical financing of investment.

4th category – An established product for which the market is expanding (a new function for it has been discovered, or it has a new use). Bank finance more likely than not.

5th category – A new product which expands the market of an established product, through a new function or new use. A bank is unlikely to provide finance.

6th category – Means of production, measurement or control involving a complete change of techniques. Refusal by the bank is probable, except perhaps after the acceptance of the means by a client enterprise, which, after tests, declares itself satisfied.

7th category – A completely new product for which no market as yet exists.

The need has not yet been defined. The classical sources of finance will certainly refuse support.

It is in the more difficult of these cases that the innovation patent and warrant could be valuable. The expertise which they have involved could prove to the bankers that financial aid is worthwhile; that is, technically, economically and commercially, the enterprise is well equipped. The venture capital firms should direct themselves to the emerging innovations in the immediate aftermath of their invention (recognised as original by the invention patent). Hitherto, politicians, following the American example and believing it necessary to complement the banking system, have praised the generalised recourse to risk capital. But, in France, the firms which finance innovation do not have the same structure, role or volume of resources as their American counterparts. They can only finance a limited number of projects. They must concentrate on second-stage activities.

Nevertheless, the financial aid which they provide may well be more important than it appears. The very fact that they are involved as a result of sound and expert investigation, can open the way to supporting funds, most frequently offered by firms outside the financial sector, which have an interest in the development of particular types of innovation. Regrettably, in numerous cases, acceptance of such new capital is refused by innovators in small- and medium-sized firms, since they fear the loss of control. This fear can be instinctive in origin. The appetite of big businesses involves integration and absorption under different guises, which can only worry those innovators who are anxious to keep their freedom of action. The same policies, however, reassure those who prefer to find an environment within which they can grow more easily.

What we have here, probably, is a problem of "the rules of the game"! A code of good conduct, with arbitration systems etc. could produce a healthy climate so that the innovator would be able to see that he will conquer the market more easily if he is affiliated to a powerful organisation, than by remaining a small isolated manufacturer. These "rules of the game" could well be linked to the creation of the new patent systems.

It would be opportune also to force on the attention of all the economic agents involved in innovation transactions, the fact that adding to firms' "own resources" is often the only viable solution. Numerous studies show that the resources of young enterprises are too limited and their debts too great for their objectives. The action of the Patent Offices in pressing this point on stockbrokers, investment clubs – and on new sources of capital, still to be developed – should be energetic.

3.3 New Institutional Arrangements?

The consequence of what has just been said is that patent offices can no longer remain just recorders of applications which are more or less rigorously examined, according to the country. They would be called upon to become active centres for promoting invention and innovation. They could follow projects from the invention stage to the actual realisation of the product, not only granting recognition at each successive step, but also acting as promoter of the innovations through the different stages of development.

As the logical consequence of this reasoned assessment, I would therefore propose:–
– *Giving up the existing methods of evaluation.*
The "inventive level" seems to me to be a byzantine construction, riddled with subjective judgment. We may accept that precise references as to each stage of development, the typology, and the end-products of innovation, as well as technological, economic and commercial criteria will be available from now on.
– *Reduction (or total elimination) of patent costs.*
Grant of a patent is a public service. It should be considered now, with Mansfield, that even if there is a "private return" from the innovation there is also a "social return" for the community. This social return is very much greater than any gains made by individuals or firms.
– *Settlement of conflict between patents should be underwritten by the State.* There should be means for arbitration of conflicts between those granted patents. So legal actions should be undertaken by the granting office, which should also be responsible for litigation concerning rights obtained in foreign countries. Here I endorse W. Kingston's point about "lèse-majesté." This applies when a country has evaluated, recognised and brought into its strategy for development, a possible innovation (by the invention patent) a potential one (by the innovation warrant) or an actualised one (by the innovation patent). For the State to undertake the burden of protecting such an innovation, is certainly in the national interest. This is equally valid whether the protection is granted to nationals or to foreigners. Infringement has become too widespread and too lucrative an activity. It must be attacked.(20)
– *Elimination of financial aid for innovation.*
This is badly distributed, and is obtained more often by skilful writers of proposals than by real innovators. It is too diffused, it is open to criticism, and it is difficult to measure its effect. (19)
Such aids to innovation would have no further justification once the arrangements envisaged to enable a country's financial institutions (banks, venture

capital institutions, stock-exchange, public procurement policies, etc.) to operate at maximum effectiveness, were established.

The reduction or elimination of the costs of obtaining and keeping a patent,
support in settling disputed claims, and suitable protection for innovation in
all its phases, would be worth far more than financial assistance. They would
probably cost the exchequer less, and they would reach their targets more
frequently. In most countries, redeployment of organisations formerly charged
with aid to innovation, around the patent office, would give it the resources it
needs in terms of expertise and evaluation.

– *Re-organisation of innovation policies around a new type of Patent Office.*
This office would become the moving force in all such policies, since only it
can be fully aware of all their diverse aspects. It could even decide on the
introduction to a country of discoveries made abroad. It should be within its
competence to say at what stage of development these findings should be
brought in; at what time, in order to attain a given objective, research should
be initiated in the country itself, and whether this should be fundamental or
applied. In many countries, such a re-organisation might even take the place of
a Research Ministry, and promises greater realism and efficiency as a result.

– *Enlarging the range of what can be patented.*
By its very nature, an Industrial Revolution brings forward ideas, activities
and products which were previously unknown. The innovation office should
be watchful for these, and should do all it can to insert them into its activities.
Two contemporary examples are biotechnology (2) and microchips. (22,25)
The mutual interaction of the threefold system (invention patent – innovation
warrant – innovation patent) carries with it a further opportunity to obtain
the widest possible "coverage" of innovation. In effect, even if economic
agents have no interest in disclosing what they are working on, during the
process of innovation, they have, in contrast, a keen desire to make known its
final result, which has to be sold widely.

The innovation patent will thus appear, to them to have an important
bearing on publicity and sales promotion. From an early stage a manufacturer
will see the value which the innovation warrant could hold for his work, by
facilitating the provision of finance for his productive investment. Moving
further "upstream" in the innovative process, he will be able to envisage what
assistance the rejuvenated invention patent could offer for his development
work, prototype manufacture etc. The organisation of technological monitoring will then be both easier and more comprehensive than at present. (21)

This will be linked inextricably with economic, commercial and financial
monitoring. The Innovation Office will be founded on a veritable Institute of
Innovation Relationships (dealing with technology and the economy). The

attention of this Institute should be focussed on long term fluctuations in the economy, the "long waves" that are linked to the flow of innovations. (16)

Another area which needs monitoring is the strategy adopted by firms, especially in relation to patents. Recent studies have shown that obtaining "blocking" patents, in order to prevent entry to defined areas of activity through delaying competition or stopping it completely, is more frequent than had been thought. (23)

The arrangement suggested by W. Kingston for shortening the duration of patents, should be a suitable weapon with which to fight these forces of stagnation. Moreover, the intensification of the sale and exchange of patents and of granting licences, would reconcile both the private and public interest, which is to reach the stage of actual exploitation as soon as possible. (24)

One final observation: There must be no question, in this reform, of developing a huge bureaucracy; co-operation between the public and the private sectors would be one of the essential conditions of this project. It is true that in numerous areas of political economy, the interests of the community and those of individuals seem to be opposed. Happily, as far as creativity and innovation are concerned, these interests are convergent. We should make the most of this fortunate circumstance.

CONCLUSION

The present time is very favourable for root-and-branch reform. The economic crisis has already lasted for 10 years. We are in the transition between the end of one long wave of innovations and the laborious beginning of a new long wave. (13,16).

Numerous sources of liquid capital are available. The stock exchange indices in several countries have been rising significantly. This movement is not due to a return to prosperity, rather is it that in narrow markets where shares are already over-valued, genuine opportunities for new investments are rare.

In making it possible to direct capital towards innovation, a modernised, enlarged and comprehensive patent system will bring Western economies to the end of a long recession and inaugurate a period of growth which should sustain them for very many years.

NOTES

1 A. PIATIER: L'innovation dans l'industrie. Les enseignements de quelques enquëtes – CPE – Centre de prospective et d'évaluation – Ministère de l'Industrie et de la Recherche. Paris, Mai 1984.

2 F.K.BEIER et J. STRAUS: Les Brevets dans une société de changement rapide des sciences et des technologies: Les inventions dans le domaine des biotechnologies – OCDE (SPT 84-12) Aout 1984.

3 A. PIATIER: Barriers to innovation (CEE Report) Frances Pinter (Publishers) London 1984.

4 B.L. BASBERG: Patents, Innovations and Technological Development in Norwegian Whaling 1880–1968 – OCDE – Science and Technology Indicators – Conference – 15–19 September 1980.

5 André TEISSIER DU CROS: La recherche d'activités et de produits nouveaux. Editions d'organisation – Paris, 1976.

6 A. PIATIER et BURTSCHY B.: L' innovation dans les PMI – Une enquëte par sondage dans 10 secteurs. Etude réalisée pour la Direction de la Prévision – Ministère des Finances Paris – Mars 1984.

7 Cf. Input: International Market Opportunities for On-line Data Base Services – Septembre 1980.

8 However, there are some references, for example, FROST and SULLIVAN : Data base Markets in Europe – October 1980.

9 J.H. LORENZI et J. TOLEDANO: cf. Les Banques de données – Le Marché de l'information automatisée – Rapport établi pour la DSTI de OCDE.

10 See also ANDERLA: L'information scientifique en 1985 – Une étude prévisionnelle des besoins et des ressources – OCDE - PARIS 1973.

11 J. de ROSNAY: Université, Industrie et Innovation – Le cas des biotechnologies – European Management Forum – Davos – Fev. I981

12 This research indicates that relationships are complex. Factors are supplied in groups, or they converge, or they are affected by "feed-back" etc.

13 A. PIATIER: "Long Waves and Industrial Revolutions" – (on the 'clustering" of major innovations). December, 1984: INTERNATIONAL INSTITUTE FOR APPLIED SYSTEMS ANALYSIS, 2361 LAXENBURG, AUSTRIA.

14 Commission des Communautés Européennes: Réunion d'experts sur les Brevets – 3 et 4 Novembre 1982 – Luxembourg – et notamment Rapport Géneral d'André PIATIER.

15 A. PIATIER: Les relations entre connaissances, brevets et innovations industrielles. OCDE (DSTI SPR – 82-48) Juin 1982.

16 Christopher FREEMAN: Long Waves in the World Economy. Butterworths – London, 1983 (especially Chapter 16): A. PIATIER: Innovation, Information and Long Term Growth.

17 Conseil national de la statistique – groupe de travail sur la technologie et l'innovation – Rapport interimaire (Mai 1983) par André PIATIER, H. DUPRAT, P. BARTOLI – Deuxieme rapport (Octobre 1984) par André PIATIER, H. DUPRAT, J.F. MINDER.

18 A. PIATIER: La mesure statistique de l'innovation. Courrier des statistiques I.N.S.E.E. – Octobre 1984.

19 A. PIATIER: L'utilisation des enquêtes pour l'évaluation des politiques d'innovation – DSTI/SPR/83-41 – OCDE Juin 1983.

20 Marie Claude CELESTE: Contrefaçon: Une industrie florissante – Jeune Afrique Economie – 22 Novembre 1984.

21 For example, I.N.P.I.: Les techniques de demain: quelles sont-elles? D'ou viennent-elles? Publication semestrielle.

22 C. et G. BERTIN et M. PINSON: Les conditions de protection de logiciel en France. Etude N° 3 – Association de Recherche Economique en Propriété Intellectuelle et transferts techniques (AREPIT) – Paris – Mars 1983.

23 J.R. EDIGHOFFER et G. BERTIN: Le Management des brevets dans les groupes multi-nationaux. Bulletin SEDEIS – 1984.

24 C. FERNANDEZ: Le marché interieur français des licences de brevets – Etude N° 7 – AREPIT – Paris 1984.

25 A good illustration of this problem is given in a "Flash" of C.P.E. of 17 December 1984, of which the following is the text (in translation): "It appears that, after painstaking efforts, legal arrangements for protecting integrated circuits have been established in the United States. The Patent Office held that Copyright was the appropriate means for protecting the work of the originators of these circuits, whereas the Copyright Office took the opposite view – that they should be protected by Patents. In the event it is the idea of rights of reproduction (Copyright) which will be used, and not that of Patenting. The proposed Law, passed by the Senate in May, has also been approved by the House of Representatives. According to this, the originator of a design for an integrated circuit will be protected against copying for 10 years." Reform ought to employ all possible forms of protection (copyright, design registration etc.) *as well as* Patents.

CHAPTER VI

Thomas Mandeville and Stuart Macdonald

INNOVATION PROTECTION VIEWED FROM AN INFORMATION PERSPECTIVE

INTRODUCTION

The proposals of Kronz and Kingston, which form the core of this book, provide one of the most original contributions to the patent system to appear in recent years. They are lucid, they are exhaustive, they are informed and they are constructive. They propose a radical alternative to the present patent system, one based on protecting innovation rather than mere invention. We are wholly in sympathy with the philosophy behind Kronz and Kingston's proposals: invention is the easy part of technological change, innovation very much more difficult.(1) Here we part company with them. While we accept that innovation is much more valuable than invention, we do not accept that innovation should, therefore, be protected. We have argued elsewhere that the patent system works only because it does not work: that its very weakness is its real strength.(2) The existing patent system, with all its ludicrous posturing, is a quaint sideshow accompanying the major drama of innovation; it can be tolerated as amateur theatricals provided by patent lawyers and bureaucrats, but their vulgar entertainment must not be allowed to interrupt the professional subtlety and artistry of the main performance. Kronz and Kingston apparently believe that, with practice and new discipline, these fringe thespians can become stars if only they are given a leading role to play, if only they concentrate on innovation rather than invention. We believe that the whole play is just too intricate and difficult for such intrusion, no matter how well-intentioned.

THE PATENT SYSTEM AND INNOVATION PROTECTION

Kingston presents an attractive, almost captivating, argument which contrasts his and Kronz's alternative system of innovation protection with the existing patent system. As the latter is so monstrously flawed, any comparison would show any alternative in an attractive light. Kingston's shining exposition, therefore, makes Kronz's system of innovation patents and his own system of innovation warrants seem dazzlingly brilliant. Most unfortunately, the display is an illusion. While innovation protection bravely casts its light on the issue central to technological change, in doing so it reflects most of the weaknesses and assumptions of the existing patent system.

The patent system is based on a simplistic – almost static – view of the way innovation happens. It assumes industry to be manufacturing industry, and from a green and pleasant vantage point seeks to regulate change beneath dark, satanic chimneys. Innovation is imagined to concern mainly the production of technical information embodied in tangible goods, and to occur largely within the firm. The process of change is seen to be encompassed effectively by the Schumpeterian notions of invention, innovation and diffusion. In this neat model, firms calculate their investment in innovation rationally on the basis of the future returns it will provide, and are discouraged from investing if they cannot appropriate the information generated by such investment. An effective patent system is seen to be one which guarantees appropriability to the inventor and thereby provides a strong incentive for firms to invest in invention. Better still – following exactly the same line of argument, with all its traditional assumptions – would be a system of property rights which guaranteed appropriability and also provided a strong incentive for firms to invest in innovation. That is what Kingston and Kronz propose, but their model is the existing patent system with nearly all its underlying and mistaken assumptions. All that is really different is the concept that innovation is more important in technological change than invention: that innovation should be thought to be appropriable in much the same way as invention and for much the same reasons, betrays just how absolutely fundamental the acquisition of property rights is imagined to be in the process of technological change.

Perhaps the best indicator of how embedded in the existing system are the new proposals, is that only the possible benefits of directly protecting innovation have been stressed. Social costs of monopolies on technological information, such as restrictions on use, higher prices, and inefficient methods of production owing to insulation from competitive pressures, have not been mentioned. The arguments have, of course, been well covered elsewhere,(3) but this does not excuse omission of acknowledgement that innovation protection must impose some social cost and cannot convey only total social benefit. All EEC governments have patent systems and thus presumably accept the view that their social benefits exceed their social costs.(4) Although the proposals of Kronz and Kingston represent an implicit challenge to this assumption, nevertheless they embark upon the relatively simple and pragmatic task of improving the effectiveness of intellectual property rights as incentives for innovation. While they labour from the perspective of the existing property rights system, our own perspective is very different and leads us to conclusions very different from theirs. We see the innovative process as complex and dynamic rather than simple and static; we see information not as an ordinary good to be protected like any other, but as a resource of primary importance in a modern economy and with its own highly peculiar economic characteristics.

INFORMATION AND THE INNOVATIVE PROCESS

The view of innovation which sees the production of technological knowledge as something that happens mainly within the isolated firm, is extremely narrow. A more realistic view embraces the concept of innovation as a social process.(5) The production of technological knowledge depends mainly on the flow of technological information among firms. It is a process whereby bits of information are gathered from a variety of sources, mostly outside the individual firm, to be assembled in new patterns within the firm. Most information is not to be found already lurking somewhere within the firm and very little is actually created there, no matter how strong its research and development department. Technology builds on technology in a cumulative manner reflecting two economic characteristics of information. Information cannot be exhausted, but its quality can be enhanced by adding new information to the existing stock. Since the cost of production of information is independent of the scale of use, it pays an industry as a whole to share it as widely as possible. Silicon Valley is outstanding in high technology industry because the participants in that industry acknowledge information as a fundamental resource and have adopted mechanisms to cater for its flow.(6) Such mechanisms include informal networks, highly mobile experts, second sourcing, and, of course, imitation.

Innovation protection as proposed by Kronz and Kingston would provide strong property rights on technological information. Those rights might well actually stifle the overall innovative process by discouraging the flow and sharing of technological information necessary to the production of innovation. The present patent system is generally too weak to have this effect; in other words, it works only because it does not work. Nelson has recently recognized just this point:

"While there certainly are some dead-weight losses associated with restrictions on the use of particular techniques, and some waste involved in the race to be first to come up with an invention or to invent around somebody else's patent, I suspect these costs are small compared with social costs that would be involved if the background knowledge to facilitate the next round of R&D effort was kept largely proprietary".(7)

Firms require strong property rights in the goods and services they sell, but not necessarily in the technological information embodied in such goods and services. Indeed, it is quite clear that in high technology industry, firms could not innovate without the technological information of others. Though the pace of innovation is less furious in other industries, it is still the case that even the largest and most self-contained of firms in the most sedate industries is dependent on information flow from beyond its own purlieu. Information flow by means of such formal mechanisms as published patent specificiations and licence agreements is likely, in all industries, to be quite inadequate to achieve innovation. Instead, firms must resort to as wide a range of information channels as possible, and – as is particularly evident in high technology industry – to personal and informal means of information transfer. The high information intensity of high technology industry makes personal information networks essential and information must frequently travel "on the hoof" if it is to provide full and timely input to a firm's innovation efforts.(8) But high technology firms are different only in the urgency with which they gather vast quantities of information and in their explicit recognition of their dependence on external information sources. To some degree, all innovative firms in all industries are in precisely this position. Technological information, then, would appear to be one area in which it is in the common interest that property rights remain vague, weak and ill-defined.

High technology industries are no longer just a curious contrast to traditional smokestack industries. Some, such as semiconductors, computers, communications and office equipment are now important industries in their own right. More important, though, is the rapidly developing dependence of all other industries on the products and processes of high technology industries, a dependence which means those other industries must increasingly exhibit the characteristics of high technology industries.(9) In addition to high information intensity, these characteristics include an extremely rapid pace of technological change, intense competition, interaction among clustered firms, heavy and regular expenditure on innovative activity, and an important role for both small firms and individuals.(10)

Heavy expenditure on the resources required for innovation is a continuing and major part of a high technology firm's costs. It is not occasional expenditure incurred only when benefits are sure to exceed costs. Failure, of course, is common, but success can be spectacular, particularly for new, small firms started by risk-accepting entrepreneurs. Capability market power in the form of both technical and marketing expertise diminishes the problem of appropriability, although persuasive market power can also be important. Innovation protection would neither bring forth more innovative effort from successful high technology firms, nor could it, for long, instil new vigour in firms that have failed.

The Schumpeterian notion of innovative activity consisting of invention, innovation and diffusion may be crucial in the justification for, and administration of, protection of innovation, but this model no longer adequately represents reality.(11) The model assumes a linearity which is convenient but almost certainly unjustified. This linearity is explicit in Kronz's model, in which,

"Just as protection does not apply to the invention phase, which precedes the innovation phase, neither does it apply to the diffusion phase, which follows it".(12)

While there probably is some linearity in the process of technological change, it is far from clear just where the impetus starts and what direction it takes. Invention, it seems, may spring from innovation, development from marketing information, applied research from production problems.(13) Kronz and Kingston's propositions seem to be based on the assumption that innovation is the end of one process and diffusion the start of another, as if innovation – first commercial exploitation – marked the point by which technological crystals had formed and were ready for immutable distribution. If this were so, an innovation could be regarded simply as a complete technological module requiring only to be plugged into other organisations, other economies and other cultures.

In reality, technology is the totality of information which allows things to be done, and total information is unlikely to arrive as a crystallised package from the conventional research and development process undertaken by the innovating firm. What does emerge from that process is information which must be supplemented by other information before things can be done. Some of this other information is embodied in associated technologies, some is analogous to the software necessary to render computer hardware operational, but much of it is information required to effect complementary organisational change. Much of the usefulness of new technology depends on the way it is used, on information derived from, and applied by the users themselves. So crucial is this user information reckoned to be that a process of horizontal innovation has recently been postulated in which users themselves are largely responsible for the gathering and transferring of technological information.(14)

In this situation it is hard to see much justification for strong property rights on only the tangible bit of embodied technological information emerging from a particular innovating firm. There is often little to the innovation process that is tangible and a doctrine that "only tangible objects can receive innovation protection"(15) leads to some extraordinary contortions: For example, in the innovation patent of Kronz,

"The "settings", "readings", or "timings" of all the interacting components of the apparatus would be given, as well as their mode of interacting, and the inputs and outputs of the operation would be given in terms of energy and materials".(16)

Innovation warrants, too, are a clumsy and largely inappropriate tool that could very easily, and very seriously, interfere with the delicate working of the innovative process.

THE INFORMATION ECONOMY AND ECONOMIC CHARACTERISTICS OF INFORMATION

Since Machlup first drew systematic attention to the existence of information activities,(17) evidence has accumulated – in the United States,(18) across the OECD,(19) and most recently in Queensland(20) – demonstrating the significance of information in a modern economy, and tracing the growth of an information economy. More and more people in the labour force are being employed to handle the production, processing, distribution, storage and consumption of information. These activities comprise the information sector, and include most office activities, the media, computers and telecommunications, and, of course, research and development. Since the 1950s, with the advent of the computer and the growing capacity of intelligent electronics to handle information, the information sector has been growing rapidly in all advanced economies, and now accounts for between one third and one half of the labour force in these countries.

Society would appear to be in transition from an industrial era, with its emphasis on factories and transport, to an information era, with its emphasis on offices and communications. Information is becoming the prime resource. Its production, use and communication assumes the dominance in the information era that mass production of goods and services assumed in the industrial era. In this early phase of the information economy, the dominance of one form of information in economic activity, namely technology, is now generally recognised. Denison's work has shown that advances in knowledge affect economic growth to a greater extent than growth in capital and labour inputs, and governments place high priority on science and technology policy to enhance the international competitiveness of their industries.(21)

As information activities grow in importance, the economic characteristics of information will increasingly influence economic activity and the nature of institutions governing it. Information is intangible, inexhaustible, and indivisible. The implications of these characteristics are only just beginning to be understood: costs of producing information are independent of the scale on which it is used, information has many of the characterstics of capital, information is accumulative and synergistic, information is often inherently a public good.(22) These features bring both advantages in sharing and co-operating, but also difficulties in appropriating information for exchange in formal markets. Mechanisms for converting information from an inclusive public good to an exclusive private good – packaging it in tangible form, laws of contract, trade secrets, patents and copyright – become increasingly imperfect as the information era ripens. Basically, such mechanisms are out of step with the economic characteristics of information. The concept of strong property rights in information is becoming increasingly inappropriate. In his dissenting statement to the recent Australian Government inquiry into the patent system, Lamberton emphasised the need for policy to avoid "the restrictive consequences and additional social costs that can arise if the scope of the patent system is extended unnecessarily in the development of the information economy".(23) This is not merely a matter of the existing patent system being unsuited to such high technology industries as information technology, though patently unsuited it is.(24) Much more important is the recognition that high technology industry is merely splendidly illustrative of the dependence on information of a great deal of other economic activity, and that no formal system of information appropriation – be it directed towards invention or innovation – is well suited to the requirements of an information economy.

INNOVATION PROTECTION IN PERSPECTIVE

In our view, protection of innovation might have the opposite of its intended effect – it might well stifle the innovative process. An example will illustrate the argument. Suppose Firm A devises a new product, as Apple did a few years ago with one of the first personal computers. As this product begins to sell, Firm B is induced to enter the market and does so mainly by imitating, but perhaps also by improving slightly on the product or by introducing a small variation which might appeal to other market segments. Similarly, Firms C, D, E and F also enter. Small improvements and ready access to the information which produced them create the potential for product change. Firm A may now introduce a radically new version which builds on the cumulative improvements of others and also adds new features of its own. As the market develops, the pace of change quickens, reflecting technology synergistically building on technology. Firms have to be innovative and quick on their feet, although they all enjoy some natural protection via capability and persuasive market power. Some fail; indeed, failure is an integral part of the process. Occasionally the failures are dramatic, as in the cases of Atari and Texas Instruments in the personal computer market. But the product continually changes and improves as technological information flows freely among firms which are not prevented from using the technological information of others. The pressure of competitive rivalry is the prime driving force and patent protection in its present form is not sufficiently strong to prevent imitation.(25)

Now consider this same scenario with innovation patents or warrants firmly in place. Firm A devises a new product, such as a personal computer. Kronz's "option claim" system does permit competitors to use the information contained in it under license, but Kingston's innovation warrant provides total protection, and insulation from competitors. Even close substitutes developed by other firms will be outlawed by Kingston's doctrine of "commercial equivalence",(26) and so there will be no development of the product outside Firm A unless it is achieved amidst a tangle of litigation and onerous royalty arrangements. If Firm A does not improve its product much, for the term of the warrant there will be little development anywhere. Indeed, the product is unlikely to improve much because Firm A is protected by its innovation warrant from competition and because of the limited resources possessed by any single firm in an industry. A single firm is unlikely to be able to improve a technology to the same extent as a vigorous cluster of competing firms. Perhaps this delay is intended; Kingston certainly sees a possible role of his innovation office in the use of its powers to slow the rate of technological change in order to both reduce and spread its human costs.(27)Kingston admits that protection in the form of import controls ossifies industry,(28) but apparently cannot see protection of innovation having a similar effect. When he imagines that innovation warrants will temper the rate of technological change,(29) he argues that new warrant holders should be asked to compensate old warrant holders in a similar area. Even here, though, we fear that Kingston has underestimated the full implications of this compensation proposal. If the new always had to bear all the adjustment costs of the old in the process of technological and structural change, such an albatross would be an enormously powerful disincentive – not an incentive at all – to investment in innovation.

A key rationale for the innovation protection concept, according to Kingston, is that it will stimulate innovation by small firms. This variation on the infant-industry argument perceives small firms as requiring protection to balance the capability and persuasive market power of large firms. Kingston asserts that,

"When capability market power underwrites the bulk of innovation, the tendency...is for growth to be reflected in the increasing size of existing firms".(30)

In fact, small firms do possess capability market power in the form of human capital, as is reflected in their major role in high technology.(31) In some industries, where there is an association between entrepreneurialism and innovation, it is small rather than large firms which are dynamic and innovative. In these circumstances, it is nonsensical for Kingston to assert that innovation protection will allow the small firm to be

"... ... brought into dynamic contact with the technology of the larger firm, which is likely to be more advanced than its own".(32)

Kingston thinks it desirable that firms direct their creative energy to innovation rather than to solving the riddles of the present patent system,(33) an opinion with which we heartily concur. However, his only perception of compliance costs associated with patents is in connection with patentees being responsible for protecting their patent rights. Thus he would shift the burden of prosecuting infringement to innovation office bureaucrats. Yet patent litigation is only a relatively small part of the compliance costs that the existing patent system forces on users and industry as a whole. For users there is the time-consuming and expensive procedure of applying for a patent. At least in the present system this process ends on submission of a complete specification. The innovation patent concept of "step by step" disclosure or the innovation warrant's "staging" of investment(34) would guarantee a long sequence of expensive transactions between firms and the innovation office. Moreover, in the case of the innovation warrant (though not, we accept, in that of Kronz's innovation patent) its very strength would force all innovators into the system, and also compel all interested parties to become intensely and expensively involved in the opposition proceedings. None of these activities, like patent litigation, has anything to do with innovation *per se*.

At present, Australian evidence suggests that patent specifications are not a particularly important source of technological information.(35) To the extent that industry does use them, the motive is mainly the legalistic one of checking on potential patent infringement. Time spent checking that new ways of doing something are not infringing someone else's patent monopoly is time that could probably be spent in more productive ways. Again, the very strength of innovation warrants will force all industry into this unproductive pastime. Kingston's perception of these costs is entirely different; he sees them only as benefits:

"...monitoring of such published applications would be vastly more important to all firms than monitoring of Patent Specifications is now. This is a valuable feature of the Warrant system from the aspect of improving awareness of the state of the art amongst all types of businesses, and of stimulation of innovatory activity as a result".(36)

PRACTICAL PROBLEMS

Even the most staunch advocates of the existing patent system would not dispute that there are practical difficulties inherent in that system, most obviously those concerned with examination, enforcement and awareness of publication. It is, however, a characteristic of traditional institutions that their absurdities become more acceptable with the passage of time, even that an impractical procedure or nonsensical ceremony is required to prove lineage and respectability. The foibles of the present patent system are highly evident and are accommodated – though at a cost – by all those who would not attempt to take that system too seriously. But Kronz and Kingston's proposed system of innovation protection is brand new and is meant to be taken very seriously indeed. Consequently, the potential problems it poses could be very real indeed.

There is no opportunity in this short response to explore all the detail of all the problems likely to emerge. Examples must suffice. It is totally unclear how the innovation office bureaucracy could possibly do all that is expected of it, even with the research and examination divisions advocated by Kingston. The innovation office is to assess information on both the firm and its innovation, to control licensing, to settle a monopoly term by means of a formula that is to be constantly refined by research, and it then has to assume responsibility for the active policing and enforcement of the monopoly. In Kronz's model it would continue, in addition, its existing duties of administering the present patent system. As for the latter under the Kingston proposals, we are assured that it "would remain completely untouched by the introduction of Warrants".(37) In either case, a monstrous bureaucracy would emerge and, unfortunately, not one limited to the innovation office. The ground would be fertile for the sprouting of committees galore; there are to be boards of experts, industry is to be involved in assessment, there is to be peer review – even Chambers of Commerce are to be involved in the generation of paper. For the innovation patent,

"The intention is that this documentation will eventually grow into a description of the innovation object which is finally launched commercially, that is so comprehensive and detailed as to allow an inter-disciplinary team of average persons "skilled in the Art" to copy the innovation in every respect as soon as its protection has expired".(38)

While Kingston is ready to acknowledge that time spent by industry on litigation is unlikely to be productive, he makes no such acknowledgement for time spent providing the innovation office with reams of information. Neither is there acknowledgement that, from the point of view of the innovation office, there may be some difficulty determining the accuracy and relevance of information obtained from firms. That statements be signed by auditors is thought to be enough to overcome the problem. This is an opinion which might be refuted by many a public servant responsible for the collection of taxation from companies. Even the most basic matters present horrendous information difficulties: only a firm may be granted an innovation warrant, but when is a firm a firm, and what are its links with other firms, and who really owns the concern? Such matters, of little significance in the existing patent system, become crucial inputs to Kingston's risk matrix.

The risk matrix has two components: risk inherent in the project and risk related to the capabilities of the particlar firm. The matrix allows these factors to be considered together and produces an appropriate monopoly length. So, a small firm with a risky project would be offered a long monopoly; a large firm with a fairly safe project a very short one. The formula would be applied anonymously by the innovation office, though there would apparently be continuous fine tuning resulting from the empirical research of that office, and screening as well to determine whether the innovation was already being marketed. Among the problems posed by this extraordinary attempt to calculate the incalculable, to make uncertainty certain, are those intrinsic to determining just what resources will be required for the project and what innovative capacity the firm possesses. The first will probably not be apparent until long after innovation and the second is in no way synonymous – as the Kronz model explicitly recognises – with the sheer size of a firm.(39) This sort of problem – basically what allows some firms to be innovative when others are not – has tested the abilities of those who study such matters assiduously,(40) and it will not be solved instantly by a warrant office bureaucrat using a slide rule and Kingston's matrix. As for varying the monopoly term so generated to give effect to various science and regional policies...well, perhaps an extra gratuitous element of random uncertainty would be appropriate to programmes applied in the name of such policies, and would go totally unnoticed.

168

CONCLUSION

Our comments on the "thesis" chapters may have done them a double disservice: these comments have, for simplicity's sake, ignored many of the distinctions between the Kronz and the Kingston systems, and they have failed to acknowledge adequately the usefulness of these chapters as a discussion document rather than a final solution to what is certainly an intractable problem, and probably an insoluble one. The concept of innovation protection is likely to prove extremely seductive to governments anxious to revive their economies by encouraging innovation and stimulating high technology. Yet, as we have argued from an information perspective, innovation protection is not appropriate to the realities of high technology industry and an information economy. Innovation is a social process in which both clusters of firms working in promising areas, and the free flow of information among those firms are important elements. The view that direct protection of innovation is necessary, vastly overrates the problem of appropriability. Such protection would impede the flow of information, deter the formation of clusters developing new technologies, and place a heavy burden of compliance costs on all industry. That is not the way to breathe new life into an ailing economy.

NOTES

1 See F M Scherer, *Industrial Market Structure and Economic Performance*, Rand McNally, Chicago, 1971, p.350; Joan Cox, "Planning for technological innovation. Part 1: Investment in technology", *Long Range Planning*, 10 December 1977, pp 40–44; Sven Malmstrom, "Innovation and the economic crisis", *Skandinaviska Enskilda Banken Quarterly Review* , 3/4, 1978, pp 85–97

2 See Thomas Mandeville, *Information, Innovation and the Patent System* (forthcoming)

3 E.g., Fritz Machlup and Edith Penrose, "The patent controversy in the nineteenth century", *Journal of Economic History*, 10, 1 May 1950; Harry Johnston, "Aspects of patents and business as stimuli to innovation", *Portfolio – International Economic Perspectives*, 5, 2, 1977, p. 421.

4 See C T Taylor and Z A Silberston, *The Economic Impact of the Patent System*, Cambridge University Press, Cambridge, 1973. For another view see T D Mandeville, D M Lamberton and E J Bishop, *Economic Effects of the Australian Patent System*, Australian Government Publishing Service, Canberra, 1982.

5 See Nathan Rosenberg, *Perspectives on Technology*, Cambridge University Press, Cambridge, 1976.

6 Everett Rogers, "Information exchange and technological innovation" in Devendra Sahal (ed.), *The Transfer and Utilization of Technical Knowledge*, Lexington Books, Lexington, Mass., 1982, pp. 105–23.

7 Richard R Nelson, "The role of knowledge in R&D efficiency", *Quarterly Journal of Economics*, XCVII, 3, August 1982, p. 468.

8 Stuart Macdonald, The need to succeed", *Journal of General Management*,4, 3, 1979, pp. 36–47.

9 D K Stout, "The impact of technology on economic growth in the 1980's', *Daedalus*, 109, 1, 1980, pp. 159–67.

10 Stuart Macdonald, "High technology policy and the Silicon Valley model", *Prometheus*, 1, 2, 1983, pp. 330–49.

11 See also the discussion in Stuart Macdonald, "Technology beyond machines" in Stuart Macdonald, D McL Lamberton and Thomas Mandeville, *The Trouble with Technology*, Frances Pinter, London, and St Martin's Press, New York, 1983, pp. 26–36.

12 William Kingston, above.

13 See M Gibbons and R Johnston, "The roles of science in technological innovation", *Research Policy*, 3, 1974, pp. 220–42; Sol Encel, "Science, discovery and innovation: an Australian case history", *International Sociology of Science Journal*, 22, 1, 1970 pp 42–53.

14 Eric von Hippel, "The Dominant Role of the User in the Scientific Instrument Innovation Process" *Research Policy* 5, (1976) 212–239.D Leonard Barton and E Rogers, *Horizontal Diffusion of Innovations:An Alternative Paradigm to the Classical Diffusion Model*, Working Paper No. 1214, Sloan School of Management, MIT, 1981.

15 Kingston, above.

16 *Ibid*.

17 Fritz Machlup. *The Production and Distribution of Knowledge in the United States*, Princeton University Press, Princeton, 1962.

18 M U Porat, *The Information Economy*, Office of Telecommunications, U.S. Department of Commerce, Washington D C 1977.

19 OECD, *Information Activities, Electronics and Telecommunication Technologies*, Paris, 1981.

20 Thomas Mandeville, Stuart Macdonald, Beverley Thompson and D McL Lamberton, *Technology, Employment and the Queensland Information Economy*, Report to Department of Employment and Labour Relations, Queensland, October 1983.

21 See Clem Tisdell, "The International realpolitik of science and technology policy", *Prometheus*, 1, 1, June 1983, pp. 127–43.

22 See Kenneth J Arrow, "The economics of information" in M L Dertouzos and J Moses (eds), *The Computer Age: A Twenty-Year View*, MIT Press, 1979, pp. 307–17; D McL Lamberton, Stuart Macdonald and Thomas Mandeville, "Information and technological change – a research program in retrospect", *Greek Economic Review, forthcoming; D McL Lamberton (ed.) Economics of Information and Knowledge*, Penguin, Harmondsworth, 1971; A M Spence, "An economist's view of information", *Annual Review of Information Science and Technology*, 9, 1974, pp. 57–8.

23 D McL Lamberton, "Dissenting statement" in Industrial Property Advisory Committee, *Patents, Innovation and Competition in Australia*, Report to the Minister for Science and Technology, Canberra, August 1984, pp. 79–80. See also Mark Lawson, "Lone dissenter sees need for change in patent system", *Australian Financial Revew*, 19 November 1984, p. 38.

24 Stuart Macdonald, "Patents in perspective" in *Economic Implications of Patents in Australia*, Australian Patent Office, Canberra, 1981, pp. 21–38.

25 See E Mansfield, M Schwartz and S Wagner, "Imitation costs and patents: an empirical study", *Economic Journal*, 91, 364, 1981, pp. 907–18.

26 Kingston, above.

27 *Ibid*.

28 *Ibid*.

29 *Ibid*

30 *Ibid.*

31 E Braun and S Macdonald, *Revolution in Miniature. The History and Impact of Semiconductor Electronics*, Cambridge University Press, Cambridge, 1982.

32 Kingston, above.

33 *Ibid.*

34 *Ibid.*

35 Thomas Mandeville, "Australian use of patent information", *World Patent Information*, 5, 2, 1983, pp. 79–82.

36 Kingston, *above.*

37 *Ibid.*

38 *Ibid.*

39 Morton I Kamien and Nancy L Schwartz, *Market Structure and Innovation*, Cambridge University Press, Cambridge, 1982, Chapter 3.

40 See, for example, J Langrish *et al., Wealth from Knowledge*, Macmillan, London, 1972; Christopher Freeman, *The Economics of Industrial Innovation*, Second Edition, Frances Pinter, London 1982; J Jewkes, D Sawers and R Stillerman, *The Sources of Invention*, Norton, New York, 1969.

CHAPTER VII

Gordon Tullock

INTELLECTUAL PROPERTY

It is sensible to begin a discussion of this sort by telling the reader what my own prejudices and biases are. To begin with, I believe that a reward system for new ideas is highly desirable. We certainly do not have a surplus of such new ideas. It is by no means certain that the patent system is the optimal set of rewards in this case, but it does seem to be an effective one. There are various other ways of rewarding new and important ideas, but I am going to here confine myself to the discussion of possible innovations in the patent system.

Before turning to methods of stimulating invention and innovation, I think I should devote at least some space to pointing out that this is not a universally approved activity. Traditionally, new inventions, particularly those that save labor, have been attacked vigorously by various groups. Beginning with the labor-saving devices, those individuals who have acquired personal capital in the particular skill which is made obsolete by the new machine, are obviously annoyed by it and frequently will protest its introduction. The Luddite movement in England is, of course, a famous example and it should be said that although the Luddites failed totally in the factories, they were able to prevent the introduction of new agricultural devices for a number of decades.

This specific objection on the part of people who are injured because of the loss of human capital is no longer of any great importance. Our economy is now so diverse that the number of people who hold any particular type of special skill is usually very small and, hence, of not too much political importance. Further, most of our current "highly skilled" personnel are not actually skilled in the old sense. They are people who find it very easy to adopt new production methods. In a way, their skill is a skill at learning new techniques rather than in a given technique.

The broader antagonism to new machines by people who think that they will be unemployed or that unemployment will increase, seems to be simply the result of intellectual mistakes. Economists have been complaining about this misapprehension ever since Adam Smith pointed out that the pin factory described in the *Wealth of Nations* had only four employees although it would be necessary to hire 400 people to produce the same number of pins without the advanced technology of 1776.(1)

The basic problem here, of course, is the assumption that human wants are restricted. The cutting of the cost of producing any given item normally simply releases purchasing power to buy something else and, hence, employment can pick up there. It is, of course, conceivable that the pattern of inventions might be such as to raise returns on capital more than they raise returns on labor, but historically the reverse seems to have occurred.

There is here another problem. It is frequently said that in the short run inventions cut the demand for labor, but in the long run they may increase it. As a matter of fact, in the short run, they must of necessity increase the demand for labor because whatever the new device is it must be produced before it can be used. Thus, there will be a period in which, let us say, pins are being produced by the old method while a new pin machine is being built, and clearly the demand for labor will be higher during that period than it was before the invention of the new pin machine.

There will then follow an intermediate period in which people are laid off in the manufacture of pins (2) and they will then be absorbed providing some other items which people will be able to buy because the price of pins has fallen.

There is, in addition to this rather standard (but fortunately not very influential in practice) objection to new devices which might cause unemployment, a generally antagonistic attitude against technological progress and, in fact, frequently in favor of going back to an earlier era when there weren't so many people around and the individual people didn't have so much technology. This point of view is sometimes thought to be very modern but, as a matter of fact, a number of conservative European thinkers particularly associated with the Catholic Church in the 19th Century, promulgated this attitude. It is intriguing that there are many similarities between the number of conservative Catholics in the 19th Century, and what was called the new left in the late 60's and early 70's. Note that there are only some similarities. The two movements were very far from being identical.

This nostalgia for an older and simpler day has had some effects on modern economies. The American government, for example, has retarded development in many areas partly because higher officials have this attitude. But, in general, its effect has been – and I would anticipate will be – minor. Such advanced technologies as a cure for cancer, power sources which are both cheap and pollutant free, improved agricultural techniques to make starvation even less common than it is now, etc., are pretty widely favored. In general, the person who is opposed to technological progress turns out to be opposed to only certain things which he personally doesn't want. It is, in a way, an effort to impose his preferences on other people.

So much for arguments against technical progress. Fortunately, they have not been particularly influential over the last two centuries, a fact to which we owe our current prosperity. At the moment they are perhaps more popular than they were 25 years ago, but they still seem to be a minor current in our thought. Let us now turn to methods of stimulating invention, which is a main topic of this chapter. Basically, there are four general policies we can follow. The first is to simply leave the whole matter alone and let private individuals produce what inventions they wish. This policy, which will be discussed a bit below, can hardly be listed as a form of stimulating invention, but there are people who are in favor of it and feel that it would lead to a higher rate of progress than the other and more direct methods.

It will not have escaped the reader's attention that I, personally, am opposed to this technique of doing nothing. Nevertheless, it is very widely used in present society and before about 1700 it was standard throughout the world. Today, new inventions in applied psychology, a great many new and improved technologies which simply involve more efficient arrangement of farm crops, machines on plant floors, design of shopping centers, etc., are not patentable. In some cases commercial secrecy can be relied on, but after all one of the disadvantages of commercial secrecy is that it is secret and, hence, that other people cannot build on the discovery. Further, there are considerable costs in keeping a secret secret.

It seems likely that the basic reason that we rely on the "do nothing" technique in these areas is either that rather by accident we have never enacted legislation making it possible to use patents or that patents would, in practice, not work. Suppose, for example, a surgeon invents a new technique. He alleges someone else has infringed on his patent for that technique. The only way of finding out whether this is true is to open up the patient upon whom the second surgeon had operated and look at the scar tissue. It might also be invisible even under these circumstances. Similarly, attempting to get farmers to pay royalties on management techniques that they chose to use on spacing of crops, salesmen – on the use of new sales techniques, etc., would in practice be impossible even if we were so unwise as to pass laws making patents available in these areas. In my opinion, this leads to a much lower rate of progress in these areas, but I do not believe the patent can be used to accelerate progress in them.

This naturally leads into our second technique which is to hire people to do the research. There is, of course, no reason why we couldn't hire people to do research in these areas where the patent is not possible and indeed we do. The Bureau of Standards itself does a good deal of research of this sort, the Department of Agriculture does a very large percentage of the total amount of agricultural research done in the United States, and various research agencies, in fact, provide subsidies to develop new surgical techniques. This technique is, of course, used in many other areas, too. Clearly, in areas where the patent system will not work, if we want to motivate further inventions either the procedure of hiring people to do the reseach or our third possibility, prizes, are what we must depend on. It is also arguable that one or the other of these would be better than patents in those areas where patents are in any event possible. A third procedure is some direct reward system such as prizes for inventions, but so far as I know this has never been the major foundation of any policy for stimulating invention. Perhaps that is a mistake.

The final procedure is to provide some kind of a legal structure under which the inventor owns his discovery. Patents are, of course, a prime example of this procedure, but there are other ways in which the same thing can be done. Jack Hirschleifer once suggested that the inventor purchase all of the resources whose value would increase as a result of his invention and, hence, get the full value that way. Although this solution has the advantage that it would involve no government policy whatsoever, I fear it has little practical application.

A few words now on this general classification is, perhaps, worthwhile. Firstly, hiring, prizes, and intellectual property all involve direct financial stimulus to invention, while doing nothing, of course, does not. The prize system and the intellectual property system involve rewards given to the inventor after his invention, while the hiring of someone involves paying him before he invents. Hiring inventors and the prize system involve a direct cash outlay by whatever agency, presumably the government, is attempting to stimulate invention. The first, "do nothing", and the fourth, intellectual property, are all cases in which the reward, if any reward is in fact earned, is given to the inventor not by the government but by various private citizens who choose to purchase the product. In the case of hiring and prizes, the decision as to who will receive the reward necessarily will come from whatever agency is attempting to stimulate invention.

A final preliminary word before I begin discussing the new proposals. Research in this area has been largely theoretical and we don't have very much empirical knowledge. Indeed, Mansfield's (3) elaborate studies of the social payoff of invention are, I think, the largest bit of empirical research in the field. Further research is, I think, called for. We need much more empirical

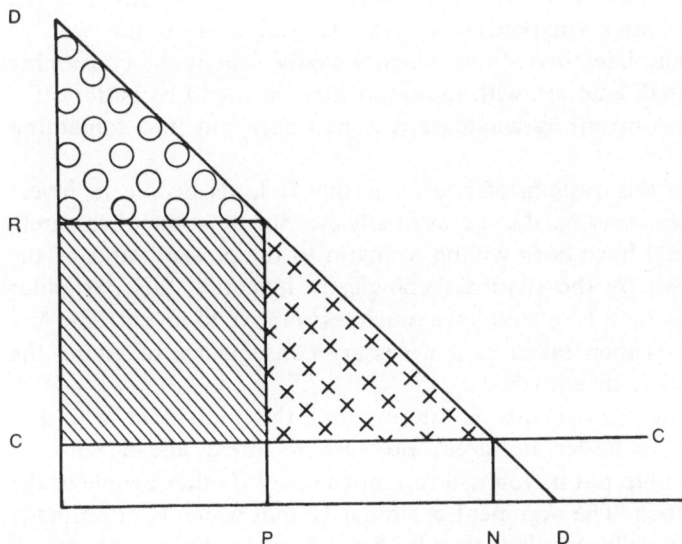

Figure 1.

data upon which to base policy decisions. Let us now return to the first of our "stimulating" policies which is simply to do nothing. Of course this policy does not, in fact, stimulate research but there are people who maintain that technical progress would be faster with this policy than with intellectual properties.(4) People who hold this point of view in essence argue that our current patent system in fact retards invention and use of new ideas and, hence, its abandonment would "stimulate" new invention. Although I do not agree, I think it is worth while to pay some attention to their argument.

Let us then look at a very simple situation shown in Figure 1. There is a demand for a particular product which can be produced at cost C. The product has been invented by someone who charges a royalty which raises the price at which it could be sold in the market.(5) P units are then produced and the owner of the patent receives the shaded area as his royalty. There is a net social loss in that there is really no cost in using the patent out to point N and, hence, socially, society as a whole would be better off if we produced N units. This social loss is shown by the X'ed triangle.

Of course, if the royalty owner were able to differentiate his royalty instead of charging everyone the same, he might permit production out to point N and take a larger royalty because of differentiation. This last proposal, however, has the disadvantage that if he can differentiate, he is apt to cut into the consumer surplus triangle (marked with O's) as well as into the welfare triangle. Thus, we may have inventions which actually benefit the consuming public only a very small amount, with the result that we would be better off if the resources put into inventing whatever it is had been put into something else.

The argument for this system, of course, is that it leads people to invest money in research. It is very hard to guess in advance the outcome of research, but the inventor would have been willing to invest to the present value of the income stream, shown by the shaded rectangle, in inventing this particular object. Without the patent, he would have much less motive to invent. The loss in the X'ed triangle is then taken as a necessary consequence of giving the inventor the appropriate incentives.

The one argument against this is simply that the individual who has received the patent is under no great pressure to press ahead with its exploitation. As Machlup put it, you will run much faster if other people in the race aren't handicapped. The argument is similar to that which is often made of monopolies. It is alleged that people fight harder under competitive situations. It may well be true, but economists usually prefer more formal arguments. After all, the monopolist is a profit maximizer, too, even if he doesn't have to work quite so hard to make at least some profit.

The question of whether hiring people to do research might not be better than use of the patent system has been somewhat obfuscated by discussion of the welfare triangle (the area marked with the x's in Figure 1) which is inevitably associated with the patent procedure. Hiring people to do research, or for that matter a set of government prizes, would not have this social cost. It would, however, have another welfare triangle, attached to the taxes used to pay for it. Whether this second welfare triangle would be larger or smaller than the welfare triangle involved in the patent is not very obvious. A number of economists, however, have said that it would most certainly be smaller in the case of taxes than in the case of patents. Even if we grant them that this is so, it is not obvious to be enough smaller so that the phenomenon would be of much importance. I don't think we can make decisions between direct government payment for research, whether in the form of organized industrial laboratories or in the form of prizes on the one hand and the patent system on the other, on the basis of this not certain and, in any event, rather small resource waste involved in the patent system.

I believe therefore, that the system is basically quite a good one even though I have not ruled out the possibility that something better might be invented. If the new idea is patentable, its creator or his assignees are given what amounts to ownership in the idea for a term of years. Thus the reward, once again assuming that the object is patentable, and that the patent can be enforced, is apt to be roughly proportional to the value of the idea.

The system has been heavily criticized on the grounds that the individual is given a monopoly. This is of course, true, but it is also true of much other property. For example, think of the fortunate people who owned the tract of land lying in the "V" between U.S. 50 and Interstate 66 west of Washington. The monopoly power given by this location has been taken advantage of by erecting Fair Oaks Mall and various subsidiary hotels, office buildings, etc.

The argument for permitting owners of real estate to acquire for themselves the full monopoly value of any fortunate location they may happen to have is normally that this leads to intelligent speculation with the result that land has a tendency to go into its highest value use, and society as a whole gains. As a matter of fact, in the United States, the zoning authorities frequently artificially create such monopolies which raise the value of the land. At least in theory, this again, can be argued for as mostly desirable because of the suppression of possible externalities generated by various types of land use. In practice, it seems likely however, that these externalities are relatively unimportant, and political influence and in some cases bribery are more important.

The same argument can be used for patents. A man buys some derelict farm land on the outskirts of Washington because he predicts that with time the changing settlement patterns will give him a considerable local monopoly power. Another man invests money in inventing an improved oven cleaner. These are very similar in their economic effect even if the technology is different. And there is no political influence nor bribery in the patent case.

So much for my basic feeling that the patent system is desirable. As a second aspect of my general feeling here, I would like to have it expanded to as many kinds of intellectual property as possible. It seems to me that we should attempt to get new and desirable ideas in all fields, not just those of mechanical invention. There are of course, other incentive systems, the publish or perish rule in academic life for example. It seems to me that the patent system is the best so far proposed, and I would like to have it as widespread as possible. Our stock of knowledge is in some ways, the most important capital that we have in society. It is a very special form of capital, but nevertheless, expanding it is highly desirable and we should invest in it up to the point where the marginal return on the last dollar invested is equal to one dollar. Mansfield's studies indicate that even in the formal patent system today, we have not reached that goal. Surely in areas where the patent does not apply, we are even farther behind.

In considering the problem, how to stimulate invention, it sould be said that a great many inventions are not particularly scientific. The improved axe which is now sweeping the market (6) was the result of non-technically trained men coming to the conclusion that there should be a better way to split wood. The device could have been invented and manufactured in ancient Rome. This is hardly scientific and yet as an invention it is probably more important than a great many things turned out by the duPont laboratories. There is, of course, also the problem of stimulating the actual application of an invention. It does us little good to know now, historically, that there was a professor in England who was a contemporary of Galileo and appears to have independently discovered, sometime before Galileo, many of the things Galileo discovered, because he simply entered them in his notes and they have only recently been dug out of the library. We want not only that there be inventions but that there be innovation, i.e. the actual use of them. Further, and here I come to the main theme of this chapter, we would like innovations which improve our economy even if no invention is involved. When Marriott decided that it would be a useful addition to his restaurant to have a parking lot with special waiters called car hops who served meals on little trays which were designed so that they could be attached to car windows, he was improving the organisation of part of our economy. The idea was not original with him. Note, by the way, that if he had originated the idea, even though it was an important innovation in the 1920's, he would not have been able to get a patent on it under our present institutions. My own proposal in the past has been simply to extend the intellectual property which we call a patent to as many fields as possible. Currently, the exact border between patentable and unpatentable objects is a somewhat irregular one and is frequently changed by court decisions.

If we look over history, we observe fairly frequent changes in what can and cannot be patented. Computer programs can and then cannot, and then can be patented. Exactly what type of chemical operations can be patented has been changed. Further, for a period of almost 20 years the Supreme Court never upheld a single patent, and then began upholding with almost monotonous regularity every patent. A particularly striking example, albeit only in a related field, has been the erosion of the value of copyright through the invention of various electronic copying devices. These devices, at one time looked as if they might literally abolish ordinary printing. Congress has now enacted legislation which makes that unlikely but they have not yet succeeded in greatly reducing the amount of unauthorized copying that goes on. All that the present legislation does is make it impossible for well organized companies to engage in pirating copyrighted material. An immense amount is done privately in a rather ill organized way in libraries and in other places where Xerox machines are available. Here, again, the strength of the monopoly protection has changed considerably.

There are, as I have mentioned above, a number of areas where the patent system will not work. We should, however, be using it everywhere it will work and, hence, laws should be changed to give it the widest possible scope. Kronz and Kingston now claim to have found a number of areas in which it would be possible to have a patent where currently it is not legally possible. There would be, in general, no harm if legally it were possible to patent things which under present practice cannot be protected. My special surgical technique mentioned above would be an example. One would anticipate that very few successful infringement suits would be brought under these conditions, but drawing the patent law very broadly to cover all sorts of contingencies and then letting individual inventors discover that it really doesn't give them much protection in some areas is better than drawing it in such a way that potentially patentable inventions are not covered.

As can be deduced from the above, I of course welcome Kronz and Kingston's proposals to apply protection to innovations which are not covered by the present patent law. The only criticisms I have are purely technical. Unfortunately, they are real problems.

An important aspect of the innovation patent or warrant is the extension of property right in new ideas to a number of other fields. Kronz lists a number of areas where in my opinion, clearly we should have such protection.

Kingston, who does not provide such a formal list, clearly has a number of very similar areas where innovation protection could be extended. It can be seen here that I am not complaining about the basic desire to have this innovation protection if it is thought of as an addition to the patent scheme. I am objecting to some of the arguments used to support it. I do indeed have some objections to some of the details of the innovation patent or warrant as suggested, but the proposals could be corrected by eliminating the areas where I object, without basically damaging the new system very much.

There are however certain problems with the existing patent system, not in terms of procedure, but in terms of substance which I think could be corrected. They are relevant here because the same general type of problem could arise under either Kronz's innovation patent or Kingston's innovation warrant. I shall discuss them in the context of patent law rather than in their own terms. This is because my comments indeed form part of a proposal for improving the patent law which I presented in response to a commission from the United States Patent Office. The same general ideas however, would apply to direct protection of innovation.

Basically, I argue that patents are both too narrow and too broad in their coverage. This sounds paradoxical, but, of course, they are too narrow in one area and too broad in another. Let me begin with an area which is too narrow and talk about what is called "inventing around".

Suppose an individual invents some object for which he estimates the profit maximizing royalty is $5.00 and at $5.00 he will sell a million units. The resource investment in order to make the invention is $4,500,000. It is in the social interest that he invest the $4,500,000 and collect his $5,000,000 and it is also to his interest.

Suppose, however, that after having invented this invention, someone – let us temporarily assume that it is someone else – discovers invention B. Invention B performs somewhat the same services as invention A, but not quite as well. Let us suppose that if invention B is put on the market it will be able to sell at a royalty of $3 a unit, 500,000 copies, so that the inventor would be willing to invest up to $1,200,000 in inventing it. When it appears on the market, however, invention A suddenly has a competitor. The best thing that the inventor of invention A can do under these circumstances is to cut his price to $4 (his device is, after all, better than B) and even so he will only sell 600,000 items. His initial $4,500,000 investment now leaves him losing money on the invention as a whole. Note that the inventor of the original device would also have lost money in this particular case if he himself had been the person who invested the $1,200,000 to produce invention B and then simply suppressed it.

As long as the first of these two scenarios is legally possible, no one will invest $4,500,000 in invention A. At most, an invention which is expected to bring in $5 million in royalties would attract only a million or so in direct inventive effort because the owner of the invention would realize he would have to do a great deal of protective patenting around it, and the cost of that would have to be charged against the original invention. This is the area in which the current patent coverage is too narrow.

Remedies for this problem are not easy, but I think at least a partial remedy is not impossible. We should begin by noting that sometimes the first invention in a given field is not the best. A drug company discovers a particular chemical which will cure some disease. Both its own laboratories and all other drug companies and laboratories now begin trying different, but nevertheless similar, chemicals on that disease and also on diseases which seem to be somehow related. The usual outcome of this process is the discovery of another chemical which works better than the original one and, normally also, that somewhere in this mass of research chemicals curing other diseases will be produced.

What we would really like in this case, although I am by no means convinced that it is possible, would be that the original innovator who produced what now becomes a whole family of chemicals should be more highly rewarded than the people who followed in his tracks even if those people did make very genuine improvements in his discovery. The only objection to doing this is administrative, and in the new proposals, Kronz's distinction between his "copy" and "option" claims seems to me to be attempting to move in this direction. I would suggest some careful investigation of this problem. Prospects of success are not very good, but the payoff would be large enough so that a sizeable resource investment would be desirable even if the odds were against it.

The problem of pure "inventing around", however, is somewhat simpler. If we consider only those cases in which Company A is contemplating investing resources to invent device A, but it realizes that if it does so there is a real prospect of someone inventing device B, which is not as good as device A but which nevertheless would compete with it, then his motives for invention are reduced. This is, as I have pointed out above, true even if Company A, itself, proposes to invent device B.

The problem here is the new device which does the job of A either not quite as well or possibly just as well, but in any event not better. I would like to, as I have mentioned above, deal with the case in which invention A suggests invention B but invention B is better, but here I will confine myself to those cases where invention B is not better.

It seems to me that competent technicians would, in many cases, be able to detect this kind of work simply by examining invention A and invention B. Extension of the law so that such cases would be counted as infringements even though device B is enough different from device A so that it is presently regarded not to infringe the patent seems to me a real possibility. I would assume that the people making the decision would make errors, as they do today, but in a stochastic sense this should increase the payoff to research for genuine innovations and reduce the current waste (from the social standpoint) of resources in either inventing around or in protective patenting.

Note that this procedure works administratively only because it is confined to new inventions which are actually poorer than the invention with which we started. This would involve a public official making a decision as to the economic value of the new as opposed to the original invention and no doubt errors would be made. It is possible to think up some detailed rules which would reduce the number of errors, but basically we would have to depend on the stochastic characteristics of the system rather than on being right on individual cases.

Lawyers for invention B would no doubt claim that although not as good as invention A in general, it was better for some special uses. In some cases this claim would be true. Once again, we would depend on the technicians, but it should be possible for the technicians, when this claim is made, to at least confine invention B to the special use in which it is better.

It is indeed a point in favor of the Kronz – Kingston proposals that it would be intrinsically easier to make decisions of this sort if we were dealing with innovations rather than traditional inventions. This would be because a need for actual market entry in order to get an innovation patent or warrant means that we would have real life customers whose behavior would be observed. This does not solve the technical problem, but it does make it somewhat easier.

Now let us consider the case in which the present patent is too broad. Suppose that the telephone is invented and then someone not employed by the Bell Telephone Company – "Sun Dial" – invents the dial telephone. The patents still in existence on the telephone and various parts of its auxiliary equipment held by the Bell Company make this invention very nearly worthless. Basically, what we want, in this case, is that the two sets of patents be combined.

Such a combination can come about in two ways. Firstly, Sun Dial and Bell can make a deal. The holder of the original patent, however, is in a very good bargaining position here because he can, to all intents and purposes, keep the other patent from being applied. This being so, presumably outsiders, i.e. people not employed by the company that holds the basic patent, probably have less than optimum incentives to produce such inventions.

The second way in which the two ideas may come together would be, of course, if the original company had a laboratory looking for improvements and, in fact, found this particular improvement. There are two limitations on this. The first of which is, of course, that the technicians in its laboratory are only a small part of the world's technicians and, hence, there is a reasonable chance that the invention will be made elsewhere. The second is that the protection given by the Bell Company's ability to prevent other people from making use of this invention, if it is made, means that Bell is not motivated to work as hard as it should on such improvements.

Thus, in this area the present system gives an individual inventor or company holding a group of inventions a degree of protection which is unduly broad. Once again, if we had some way of allocating the costs of related inventions among them, this would be an ideal place to apply it. The original inventor of the telephone, in a real sense, should receive something on every single invention for improvement of the telephone even if they are made by someone else. On the other hand, he should not be in a position where he can impose very severe bargaining costs on anyone who invents an improvement in his technology.

Here, again, I made a proposal which will help, but certainly not cure the problem, which is simply that if an inventor alleges that this kind of problem has arisen, i.e. the holder of some previous invention is blocking the use of his invention in an effort to extort extremely high fees, then technically trained judges should look at the problem and if they come to the concusion that this is true, impose either a fixed royalty scheme or a forced merger on the two parties with the terms set by the Court. This would eliminate the very expensive bargaining transaction costs that now dominate this particular type of reaction and interaction between patent holders.

Once more, I do not think that this procedure is a cure all, and would anticipate that many errors would be made. Stochastically, however, I would think that it would improve the efficiency of the patent system and, hence, attract more resources into inventions.

Here again, this type of thing would be easier with innovation patents and warrants than with traditional patents because of the existence of actual experience which would make it easier, although not of course very easy, to assess the actual value of both the original innovation and the improvement.

There is here, an even more important problem. Any litigation connected with enforcing a patent, necessarily turns on the problem of whether that patent has or has not been infringed. The judge, or for that matter, a juryman who is asked to decide whether a particular microchip circuit is a mildly modified copy of a patented circuit, or different enough so it is not covered by the same patent, normally would have absolutely no capacity to make an intelligent decision. Consider also, the bacteria which have been carefully designed by a geneotech company to produce a given chemical, and then a set of germs introduced by another chemical company. Suppose there is a difference of opinion between the expert witnesses as to whether or not the second is a close copy and hence in violation of the patent of the first.

This is of course, not entirely a new problem. The three-element vacuum tube was invented by Lee DeForrest, who made a business of infringing patents in the then infant electronics field. The third element was put in in order to get around a patent on a two element vacuum tube. DeForrest apparently did not originally realize that he had made an immense, indeed revolutionary, improvement. The full implications of his idea were realized in other laboratories years before he caught on.

We must begin by admitting that this problem is indeed a difficult one. But there is a partial solution which is to remove decisions on such matters entirely from people trained in legal obfuscation, and Kronz is very close to this in his point 14 in *The New Type of Patent Office*, Chapter II. As a slight modification of his ideas, I would suggest that we set up a set of Courts with special jurisdiction, the jurisdiction being – for example – patents in chemistry, with a Ph.D. in Chemistry as the Judge. All decisions as to whether patents are valid or not would be made by technicians who understand what is being patented and its impact on the general development of technology far better than a legally trained Judge ever can.

This would of course, require that we pay the people who made these decisions much more than present day judges because the kind of technical expertise involved in truly high tech inventions is not cheap. I see no reason why this should be undesirable. I should also say however, that I'm quite positive that these highly paid experts would make mistakes. It is very hard to define exactly what you mean by the scope of an invention or innovation. How close must the resemblance between two ideas or machines be for one to be an infringement on the other? It is intrinsically an arbitrary decision, and there is no way we can make up rules for it because of the fact that we are talking about machines that have not yet been invented and we have no idea what they will look like. Further, it is not at all obvious that we could even invent language which would mean the same in geneotech and such straight-forward mechanical improvements as the famous self adjusting wrench. (7) Indeed, it is not obvious that we even know what is meant by the word "same" in this context. But to say that we cannot do it perfectly does not mean that we cannot do much better than now. The decisions made by technical experts would clearly be better than those made by people whose training is only in law. The United States court of Patent Appeals now has engineers as members. This is an improvement although throwing the legally trained people off completely would be better. Nevertheless, these engineers are rarely, and then only by coincidence, experts in the particular field covered by a particular patent under review.

I suppose it will be necessary to set up some kind of an appellate procedure, and therefore I would suggest a Board of three judges (all engineers) be set up to hear appeals from these initial courts. I should say, however, that personally I would be delighted to leave the situation in the hands of the court of first instance.

One criticism of the above scheme might be that I simply leave the lawyers out. This was deliberate. The basic problems in patent law are technical. Each invention raises somewhat different technical problems. There are a great many legal difficulties which have come from the present system of leaving the matter up to judges. There is, however, no strong reason for having such a complicated patent law. A much simpler law would be much more useful to potential inventors and we would not need lawyers to implement it.

It is, of course, certainly true that no law code, no matter how simple nor how complicated, will provide an answer to every problem that comes up. Any court is occasionally, or sometimes very frequently, confronted with a need to deal with a problem in which the existing law is not clear. There is, however, no reason to believe that if a law is not clear any particular outcome is better than any other. The other problem is of course, the immense cost of litigation. Still, the basic problem is to improve the enforcement, to cut the costs of litigation, and in these ways, make the patent system work better. (8)

So far, I have been very kind to the Kronz-Kingston proposals. The need to protect capital investments in a new project, I agree with. I would support the adjustable time scheme that is proposed by both Kronz and Kingston, in fact I prefer the Kingston one, athough I don't feel strongly about this. In general, I have very little idea as to how long the time period for which a patent would be valid, ought to be. Thus I am not objecting, but rather mildly approving. I now propose to turn to several areas where I think the new ideas, and those of Kingston in particular, are defective.

Frankly, it does not seem to me that these innovations (although I welcome them) deal with the most important problem we face in present day patents.

The patent system is not working well today, and in my opinion the reason that it is not working well is simply that it is not properly enforced. In a way, my principal criticism of Kingston's argument is that he is turning attention away from what I think is the most important problem.

This of course, does not imply that his suggestions are themselves wrong, but that the emphasis in my opinion is misplaced. Further the problems which now make it difficult to run the patent system will, I think, also make it difficult to run his system in particular. An outline of these problems thus is in order here.

Let me begin with the United States as the area which I know best. I have already said that the American courts were for a long period, roughly from 1947 to 1967, quite unsympathetic to patents as an institution. They tended to attempt to find ways around the patent law because of its monopoly aspect. This led to two phenomena, the first of which was a genuine restriction of the size of the monopoly, and the second was making it less likely that anybody attempting to enforce the patent would succeed. Here another defect of American institutions comes in. American law suits are themselves, extremely expensive. A recent empirical investigation of the cost of a law suit in the United States, says that litigation "pays" in that "...the parties often secure monetary results that exceed the fees they pay lawyers..."(9) Clearly the argument for entering into an enforcement proceeding on the assumption that if you win, the only thing you can be confident of is that the damages you receive will be larger than your fees, is better than nothing, but it is not very encouraging.

The general attitude of the courts has to some extent changed, and they are no longer so completely unsympathetic to patents as they were. Nevertheless, they do not seem anywhere near as sympathetic as I would like.

I've had these things to say about the patent system in connection with discussion of the innovation proposals for two reasons. The first of these is that the same problems in possibly more difficult form, will turn up in connection with efforts to protect innovation that turn up in efforts to protect patents. Second, however, is that many of the reasons for feeling the patent system is inefficient, are simply due to these cases of bad enforcement. Why do we see most inventors attempting to protect their invention by in part, keeping back part of the information, usually called know-how? Clearly, because they know the patent system by and of itself is not adequate protection. Why do we have a law of industrial secrecy?

Again, although it does cover areas other than new inventions, an important aspect of it is the problem of bad protection of patents. As a result of this bad protection, and the need to use secrecy at least in part to obtain protection, inventors or the invention laboratories of large companies must invest considerable resources in various security procedures intended to give protection that would in most cases be better obtainable through the patent law. The same is true of those other protections of innovations that Kingston discusses, i.e., the simple effort to develop a certain amount of monopoly by such things as specialized manufacturing institutions, and what is legally called good will, but what he more accurately refers to as persuasion market power. There is also a category that he doesn't discuss very much, getting your government to give you direct protection in some such way as establishing a quota for imports, rearranging the environmental protection laws so that potential competitors can't compete, etc.

Now all of these things would to some extent exist even if the patent law was well run. They would however, be less important and consume fewer resources. It is clear, that if the patent law ran better, then there would be weaker arguments for the various innovation protections discussed herein. This does not mean that there would be zero argument for them. Since neither Kingston nor Kronz actually proposes abolishing the patent law, it may seem that defense of the patent law against their criticism is beside the point. Kingston in particular, is actually enthusiastic about retaining the patent law so that his new innovation protection agency will have some competition. It seems to me however, that their criticism directed at the patent law should more accurately be taken as grounds for reforming it, than as arguments for their innovation protection. The innovation protection I believe can stand on its own.

Of course, if it is thought that politically it is impossible to reform the patent law, an attitude which I am fairly certain Kingston takes and which Kronz's sort of implies, but that it is politically possible to introduce the kind of innovation protection they have here that would be a good argument. My own personal feeling however, is that it is easier politically to reform the patent law. This is particularly true since reforms of the administration, which are what I want, would essentially be a national matter. Each individual country could make the reforms by itself.

To repeat what I was saying above, the legal situation is nowhere near as easy as Kingston, in particular, implied. Some one proposes an innovation. The decision as to whether it is or is not a true innovation is probably not any easier to make than the decision as to whether a proposed patent is actually an improvement. Indeed Kingston has quite an elaborate procedure under which people first put in a proposal for an innovation and there are then arrangements for people to object, etc.

It is true of course, that the number of innovation patents or warrants applied for would be vastly lower than the number of patents, and hence there would appear to be a saving here. The present system however, wastes a great deal of resources in having patent examiners examine patents which will never be applied. The appropriate procedure is to permit anyone who wishes to file a patent of any sort. It would be examined by anybody only if at some later point in time the person who had filed it alleged someone else was infringing it. Thus, only those cases where actual manufacture or application is attempted would there be any attention to whether or not the device was in fact a new departure.(10)

The infringement of an innovation patent or warrant would appear to be just as difficult a problem as an infringement of a patent. In both cases, the question is whether a new product which is never absolutely identical is so close that it can be regarded as an infringement. If we turn to the self adjusting wrench, the original invention only adjusted over a certain range of sizes. Sears put two adjustable heads, one on each end of the wrench, and hence it could be used over two ranges of sizes. Was this an innovation? It clearly doubled the use of the wrench. Most people, I think would regard it as in fact an infringement, but it is not absolutely obvious. When we move off into more complex devices, this becomes a difficult problem. It would, I think, be just as difficult in the innovation area as in the patent. The only advantage would be that since the innovation patents and warrants could be held only for things that were actually manufactured, the total number of potential infringements would be smaller.

Another objection I have is that it is not obvious that the requirement of actually producing the object is a desirable part of the innovation patent or warrant. Suppose we simply extended the patent system so you could get patents on all of the ideas which are to be covered by the innovation system. Further, suppose we retain the present system which is that you do not actually have to carry out the act in order to get your protection i.e., you can propose it and then hunt around for a potential customer.

The only argument that I see in favor of restricting the innovation warrant or patent to actual projects is that that is the way the patent system historically originated. It seems to me the switch from that to permitting people to simply file ideas, has been an improvement.

When I was a boy, a friend of my parents, Mr Daniel, was a professional inventor. He was not a great inventor. He made an income which permitted him to live in the upper middle class part of Rockford, Illinois, but as far as I know he never invented anything that we would regard as of major importance. He did however, make a reasonable living by designing things which he patented and then sold the patent. It seems to me that he performed a useful, if minor, social function and we should permit it to continue. At a larger level, there was a young French engineer in the early thirties who produced an improved suspension system for cars, beating all of the major automobile companies out. He patented it, sold it to General Motors (11) and for many years received a dollar every time General Motors made a car. Converting this to present day terms, he had income in excess of twenty million dollars a year.

Another French engineer, Houdry, designed and patented an improved oil refining process, beating an immense collection of well financed industrial laboratories. He became a very wealthy man on the strength of his patents.(12) More recently, Pilkington, a small family glass company in England, revolutionized the glass industry by production of float glass. This was a fairly sizeable research project for a company of their size, although Corning would have probably regarded their research expenditure on it as trivial. Without the patent system, they would surely have not have been willing to undertake that research.

There are of course, also the large collection of inventors living in what used to be called Silicon Valley, and is now beginning to be called Gene Valley. In view of the somewhat dubious patentability of their products, they have a strong tendency to start their own companies, and hence become eligible for an innovation warrant. It seems to me however, that improving the patent law, which would permit a more efficient division of labor between inventors and innovators, would be a step forward. But all of this is a defense of the patent law and Kronz and Kingston have not directly attacked it. In a way, I am pushing on an open door. Nevertheless, I think it is worth while pointing out that the patent system requires reform in and of itself, but that it is an important social institution.

To return to the actual innovation process, it seems to me as I said above, that there's no strong reason why we cannot provide the extension of the rights of patentability which in essence Kronz and Kingston want, without requiring the person receiving the patent to actually go into business before implementing his ideas. Of course Kingston's "Option period", does offer some such protection.

Thus to go back to Pilkington, they did indeed, begin selling their own float glass, but they were a rather small firm, and their own sales were a rather small part of the total effect of their activities. Suppose they hadn't tried to sell, would this have made any great difference?

The grant of an innovation patent or warrant to someone who has a new idea which could not be covered by present day patents, but which nevertheless could lead to economic improvement, does not seem, in and of itself, to require that the person to whom it is given be able instantly and immediately to begin producing. We could instead provide a protection which the person who brought the idea up could sell to someone else.

Thus suppose for example, that it was I and not Ray Kroc (of McDonalds) who thought up a new way of selling hamburgers. It doesn't seem to me necessary that I actually have the finance arranged to go ahead with the project at the time that I put in for my innovation patent or warrant. Indeed the option procedure, as described by Kingston, rather implied that I don't. I could get a sort of preliminary protection which would only become fully valid if I actually began selling hamburgers under a set of golden arches. The reason for this restriction I presume is that there are a very large number of potential innovations and it is thought to be desirable to sort of ration the number of them which receive protection. Thus it would not be possible for me to make up a list of one hundred and fifty different ways of selling hamburgers, apply for innovation patents or warrants on all of them, and thus in essence prevent everyone else from undertaking any new method of selling hamburgers.

The problem here is the same as that of preclusive patenting. A. B. Dick for many years had a substantial monopoly on mimeograph machines. The machine that they made was as good a machine as they could make, but they also held an immense collection of patents on devices which were very very close to the mimeograph machine that they produced in order to prevent a competitor from entering with a machine which was not as good as their machine, but nearly as good.

The problem is a real one in present day patents, and a real one of course with the innovation patents and warrants. One possible solution, both for patents and for the warrants, is to simply make them invalid after a short period of time if they are not used. This of course rules out one potential source of return on new ideas, i.e., that they will be adopted much later, and hence in that way would reduce the number of new ideas generated. It has other advantages which might outweigh that. What we need here is some empirical research that does not now exist.

Another method would be to have the patent office, or for that matter, the innovation office make a conscious decision as to whether or not a given patent or warrant is in fact entered for the purpose not of manufacturing something, but preventing other people from doing it. This also has its difficulties. Kingston emphasizes the lack of serious research on patent problems by the patent offices themselves. This is a more general phenomenon. As a rough rule of thumb, the government does very little research on its own functioning. It would be highly desirable that that situation be changed. Not only in patenting, but in other areas too.

My third objection also has to do with legal proceedings. I have emphasized here that our current legal procedure for enforcing patents is seriously defective. The proposal to retain our present method of enforcing and supplement it by government sponsored aid to whoever charges infringement, seems to me however, worse than nothing.

There are roughly three possible outcomes of this proposal. The first of these is simply that the government will in practice not do very much enforcing because the attorneys are busy on other matters. In a good many areas where we want a particular policy enforced, we have arranged to give private parties a motive to sue. In the United States there is an institution called the Private Attorney General, in which private citizens are given the right to sue the government if it does not act.

The so called private Attorney General type of legislation is one example, but the older provision of punitive damages in many areas also is an attempt to solve this problem. Slothful inactivity which is certainly characteristic of bureaucracies, might well lead to this proposal simply having very little effect of any sort.

The other extreme is also possible. Government agencies frequently engage in vigorous activity simply because this makes it possible for them to expand their scope. Thus we can imagine, an administration which readily responded whenever it was asked by beginning infringement proceedings against some hapless victim. This would mean that instead of the present circumstance in which infringement cases are difficult and expensive the defense against infringement would become difficult and expensive while claims of infringement would be cheap and easy. It seems likely that this would lead to just as much abuse as our present system and indeed perhaps more.

The large corporation which regularly charged infringement when anybody else did anything that came even remotely close to their activities would have very substantial monopoly power. This monopoly power would be particularly important granted the innovation warrant as opposed to the traditional patent, would actually be enforced by the state. We have too much state enforcement of monopolies right now to add this method of state expansion of monopoly power onto the present procedures for subsidizing farmers, etc., by monopoly grants.There is a third possibility. Note that I have suggested that bureaucracy might do little or nothing or it might engage in phrenetic overactivity. Another possibility would be that it in fact takes up a case only when it it convinced that there is a good claim of infringement. In other words, bureaucracy would first investigate any claim of infringement itself and then undertake a legal proceeding only if it was convinced that the claim was valid. It would also always undertake an enforcement proceeding when the claim was valid.

Under these circumstances the inventor claiming infringement, would first have to convince the bureaucracy that he had a valid claim of such infringement. The later court proceeding in a way, would be an appeal from that initial decision by the relevant bureaucracy that there was a valid infringement claim here and hence that something should be done about it. The argument that this would be advantageous has to be based on the view that the cost of convincing this prosecutorial bureaucracy of the truth (or falsity, as presumably the alleged infringer would be permitted to present his case at this stage) is cheaper than regular court proceedings. Since I am convinced that regular court proceedings (in the anglo-saxon world at least) are massively more expensive than they need to be I see no reason to doubt this. I also however, see no reason why this rather crabbed form of making the initial presentation to a group of people who will if they agree with it prosecute an infringement case, is superior to having a special patent court which uses the same proceedings as would this prosecutorial bureaucracy. It could then of course be appealed to another court as normal.

Kingston does not explain at all how the group of government agencies that is to enforce infringement cases decides whether or not a given claim of infringement is correct. If it simply accepts all such claims and begins legal action, clearly it would be subsidizing a very great expansion of the patent or innovation warrant monopoly. People who had to pay their own legal expenses because they are defendants in such action would tend to avoid situations in which there was even a scintilla of evidence that they were infringing. Thus, the current situation in which infringement suits are very difficult would be reversed. What would be difficult would be defending against infringements. I see no reason that this should be an advantage. I think that Kingston has been so impressed by the difficulty of infringement cases now that he has inadvertently gone over to the other extreme.

In general, it seems to me far better to lower the cost and improve the precision of the results of patent infringement suits than to subsidize one side or the other. Kingston does not refer to subsidizing both sides, but I gather that he would agree with me that this is in general, silly. What we need here is a reform in the enforcement procedure for patents not the creation of a new bureaucracy. Note that if we did create a bureaucracy of this sort, and it decided to ration its resources, that is it would only accept the best cases, cases where infringement seems most likely, then we would simply create a new legal proceeding in which people attempted to sell their infringement claim to this agency. Presumably it would as I've said above, be willing to listen to people who said they were not infringing and hence a subsidized suit should not be brought against them. This would lead to another legal proceeding and it is not at all obvious whether this is cheaper or more expensive and less or more accurate than our present system.

The fourth serious complaint that I have has to do with the geographic scope of the innovations. With respect to traditional patent law, and also, I believe, with respect to the innovation proposals we have a classic public goods problem. For almost all small countries, they are better off if they do not honor foreign patents.(13)

Let us assume that we are dealing with two planets that have no communication with each other and both of which have the possibility of many inventions which could be made, all of which will have a demand curve D and cost C in Figure 1. Planet A adopts a full intellectual property system with the result that every inventor of a new product will receive a payment of the shaded rectangle, and there will be a net social benefit for each new product shown by the triangle marked with the O's. Planet B, however, shown in Figure 2, does not introduce the patent system at all.

It is, nevertheless, true that when someone invents a new product in Planet B he has certain advantages over his competitors. He has a headstart, hence, there is some return on his product. We show this by drawing in, in addition to the social demand for the product, line D, another line D', which shows the demand for the new invention that the individual inventor may anticipate simply because of his headstart advantage. If the inventor makes the invention he can produce it himself at quantity P, or he can sell his technical advice to other producers for a flat fee, and in either event the product will sell for R' and he will receive the shaded rectangle as his return on his inventive effort. The social surplus, however, marked by the space with the O's, is now very much larger than it was on A.

Of course, if there were simply no cost to invention, i.e. there were as many inventions made on one planet as the other, the no-patent structure would

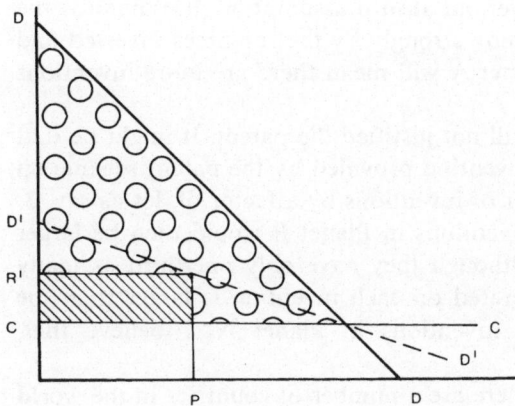

Figure 2.

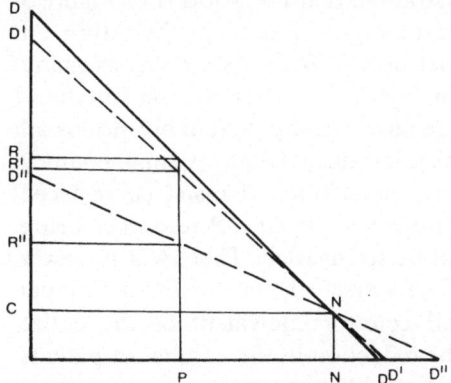

Figure 3.

obviously dominate. This is, however, an absurd assumption. Presumably, the number of inventions is affected quite strongly by the resources invested and the larger returns available on Planet A will mean there are more inventions made.

But, having said this, we have still not justified the patent. It might be that the additional encouragement to invention provided by the patent scheme on Planet A only increases the number of inventions by a factor of, let us say, 3. The social surplus generated by inventions in Planet B would then be larger than that on Planet A because, although they have only one-third as many inventions, the social surplus generated on each invention is larger than the social surplus invented on three inventions in Planet A. I believe that, normally, this would not be true.

It is sensible to point out that there are a number of countries in the world for which something rather similar to the situation shown in Figure 2 exists, and where denial of patent protection is sensible. On Figure 3, we assume that world A has many countries, some big and some small, and each individually can determine whether it will grant patent rights, not only on inventions made in its own country but on foreign inventions as well.

Assume, for the purpose of this demonstration, that the world A of Figure 3 will, in fact, be better off than the world of Figure 2 because the return on inventive effort is great enough so that, let us say, five or six times as many inventions will be made in Figure 3 as in Figure 2. Otherwise, the situations are identical, but there once again is a large universe of potential inventions all of which have demand curve D and cost C to produce. One very small country decides not to enforce patents. This lowers the effective demand curve faced by some inventor from D to D'. Thus, instead of $(R-C) \times P$ resources being invested in invention, only $(R'-C) \times P$ will be so invested. There will be fewer inventions and the world as a whole will be worse off, but for the individual small country the achieving of its small country equivalent of the entire triangle DCN more than compensates. The reduction in the number of patents as a result of this one country dropping out would have to be very large indeed before the loss to that country from reduced numbers of inventions more than counterbalanced the gain from being able to use all existing patents freely.

A large country, however, would find that if it dropped all patent support, the total number of world patents might fall quite considerably to line D'' on Figure 3. Under these circumstances, the individual inventor, being unable to patent his invention in the larger country, would only get the distance $(R''-C) \times P$ and this amount would induce fewer inventions. A large country would, of course, get a higher return for each invention that was made, but would lose more in the reduction of inventions than it would gain from being able to use what inventions had been made previously.

All of this, of course, is an argument for international agreement. If all countries agreed to honor patents, then the world demand curve for the new invention will be large if not ideal. Even if every country can gain by dropping out, every country will also gain by the world agreement. If, however, most of the countries of the world are accepting patents then an individual country which is not a member of the group can gain. The problem is analogous to that in which the company who was not a member of a cartel in its industry gains more than the cartel members.

To sum up, the defection of small countries from the patent scheme means that they don't have to pay royalties either on patents or innovations. For any single small country, the reduction in the total reward to patents and hence the reduction in the total number of patents is fairly small so the total world supply of patents will not be much reduced.

The case of this public good is like most other public goods. An agreement among all countries under which patents are worldwide is better than a situation in which none of them recognize foreign patents. Indeed, if no one recognized foreign patents it is likely that almost all inventing would occur in the United States which has the largest domestic market. Foreign countries could free ride on these inventions, but the total number of inventions would be so much reduced that they would be worse off than with worldwide patents.

This line of reasoning is, of course, the reason for the existence of the international patent convention. It is also the reason why some small countries make various efforts to get out. I find it quite astonishing that these efforts are as weak and frequently ineffective as they actually are. I suspect the actual reason for this tendency to actually enforce the patent law within national boundaries is simply that the difficulties of enforcement in the administration of patent law are so great that a very large number of innovations are covered both by patents and by secrecy. The know-how is not entered into the patent. Thus there's no point in getting out from under the patent system because infringing the patent is pretty much impossible unless you get access to certain secret information.(14)

The present situation then, with respect to patents (and so far I am not talking about innovation patents or warrants) is not a very good one. The fact that the patent is somewhat dubious in enforceability means that its owner depends in part on secrecy and secrecy has private disadvantages in that it is expensive and inconvenient, and public disadvantages in that the secrecy means that other people cannot make use of that information for further developments. Still, the system does have at least some characteristics which lead to public good development.

It is not surprising under these circumstances that small countries continuously try to weasel out of their obligations under the international patent agreement. As I said above, what is surprising is how little they do so. The geographic limits of the innovation patent or warrant provide an almost perfect excuse for such evasion. Suppose, for example, Ireland had refused to grant Pilkington protection on their glass process until it was actually used in Ireland. Further, suppose it would be willing to permit an Arab capitalist to hire a Pilkington engineer and start such a plant in Dublin. This plant would then be protected by the innovation system. Pilkington's reward for inventing the flat glass process would have gone down sharply.

Further, it might lead to a rather odd distribution of flat glass factories throughout the world. Entrepreneurs would put up factories in areas and places where they were not actually economically justified on the grounds that with a monopoly they would be justified.

There's no direct discussion of this point in either of the two proposals contained herein. I think however, that Kronz is worried about it because in talking about innovation, he includes such things as simply selling the goods into a given market as an innovation. Thus, if Pilkington had a patent on flat glass rather than on a process for making it,(15) it is possible that by simply shipping a few sheets of the stuff to Dublin, they could have acquired the innovation patent themselves.

But to repeat, although I think Kronz worries about this kind of thing, his proposal doesn't fully cover it. Early innovations in England were primarily imports of manufacturing procedures which had been developed elsewhere with, let us say, a Fleming being given a monopoly for a period in return for his agreeing to produce something or other in England.

This kind of innovation protection might then damage the worldwide reward system which we want. Perhaps special arrangements could be made in the innovation protection law or international treaty, which made this impossible, although I have no idea how it would be done. The problem would be that all of the small countries, some of which have access to extremely good legal counsel have strong motives to try to evade the rules.

I do not see any point in putting this particular temptation before them. Thus it seems to me the geographic restrictions on the innovation protection are perverse. But to repeat what I said before, this particular aspect of the public good nature of the innovation is not directly discussed by either Kingston or Kronz. It might be possible to combine the geographic restrictions of their proposals with rules which guarantee to a true innovator the full social value of his innovation.

It is fashionable these days to offer as an additional argument for any reform you wish, the statement that it will help in competition with the Japanese, that it will improve the balance of payments, and it will reduce unemployment. As a general rule, anything which improves the efficiency of the economy will indeed have some effects on all three of those areas, although the Japanese may adopt it before you do. Nevertheless, arguments of this sort are actually appeals to the common man rather than professional economists. The innovation protection idea should stand on its own merits as an improvement in efficiency. It is not likely to do very much about unemployment, balance of payments, or Japanese competitition.(16)

The American efforts to get Japan to reduce its tariff and quota protection of the inefficient parts of its economy might be largely cancelled out by innovation protection. Anyway, as good economists who know the theory of comparative advantage, we should never worry about an unfavorable balance of trade. If the Japanese export more than they import, they can either make gifts of these goods to foreigners or they can invest money abroad. In either event, they provide a benefit to foreign economies.

I would be delighted to have an extension of intellectual property into the areas which are suggested by Kronz and Kingston. It also seems to me that a basic reform in administration of the existing patent system is important, indeed more important than this extension of its protection. Since the two ideas are not inconsistent with each other however, there is no reason we cannot proceed with either or both. On the other hand, it seems to me that both Kingston and Kronz are more enthusiastic about their proposal than it warrants. Further, certain changes would have to be made in the proposal if we are not going to develop into a replacement for the protective tariff and quota system.

But having said this, I am still strongly in favor of an extension of protection in most of the areas that they recommend. The present system imposes a completely unnecessary risk on a good many entrepreneural activities. In part, the entrepreneur is a specialist in taking risk and we want him to. We do not however, want him taking risk where there is no social benefit in his doing so. With proper modifications the innovation protection provided by both the Kronz and Kingston schemes could eliminate part of this undesirable social risk. It doesn't solve all the problems, but it does solve some.

Suggestions for radical changes, and these suggestions are fairly radical, always, in so far as possible, should be tested on a small scale before we try to universalize them. There are two general ways in which the innovation patents or warrants, or for that matter, my ideas for the improvement of the patent code could be tried: one is by subject matter, and the other geographically.

To take the latter first, Ireland, for example, could introduce direct protection of innovation. We could then observe whether there was a higher rate of progress in Ireland than elsewhere. If we found innovators who lived in Tokyo who decided to try out their new ideas in Dublin rather than within Japan, that would be particularly strong evidence that the innovation protection system was working. Ideally of course, we would want a random sample of countries adopting innovation patents, innovation warrants or my reforms, and another random sample not. In the real world, the best we can hope for is that some individual countries pick up the project. Note that a large country like the United States could, if constitutional barriers were overcome, do this kind of experimenting internally, with let us say, Michigan introducing the direct protection system, then observing whether they got an exceptionally high level of innovation.

The second method would be by subject matter. For example, all changes in household products might be subject to the Kronz-Kingston system, or my reform of patents, and other areas not. Here, telling whether the experiment was more or less successful would be more difficult than with geographical tests, because there might be quite other reasons why that particular area would grow faster or slower than other areas over the period of the test. Randomization would be desirable, but quite hard to design in this case.

To sum up, we need new ideas in all fields. As an economist I think you can get people to produce things by offering rewards and hence want new ideas to be rewarded. Giving the discoverer of a new idea a property right in it, not only rewards him, but makes the reward more or less proportional to the real value of that idea. Like all property rights, it has a disadvantage that it prevents other people from using the property. If the owner of the property, intellectual or real, is unable to perfectly discriminate in the charges he makes for use of his property, then some socially desirable uses will be eliminated. This is unfortunate, but a necessary consequence of property ownership. It may possibly be more significant in the case of intellectual property, than in the case of real estate. I think the reason that it is talked about more in relation to the intellectual property than with real estate, is that since the days of Henry George, no one has really thought about unowned real estate.

The patent is one form of intellectual property and in my opinion, its administration could be greatly improved with the results that we'd have an even greater social, and for that matter, private value. The innovation patent and the innovation warrant are two proposals for increasing the scope of intellectual property, and hence to be encouraged, although it is of course necessary to look carefully at the details of these proposals. Myself, I would like to have even broader inventions of intellectual property. Intellectual property is a way of stimulating the creation of new intellectual "objects" and should be resorted to substantially anywhere we would like to have new ideas.

NOTES

1. As a matter of fact, he specifically says this was not a particularly advanced factory in terms of 1776 technology. Presumably, a really advanced factory would have gotten by with three instead of four employees.
2. Assuming the cost of pins does not fall off far enough so that the total labor input goes up.
3. "Social and Private Rates of Return from Industrial Innovations", Edwin Mansfield, John Rapoport, Anthony Romeo, Samuel Wagner, & George Beardsley, *Quarterly Journal of Economics*, Vol XCI, No.2, May 1977, pages 221-240.
4. I have heard Fritz Machlup express this opinion quite strenuously, in oral comments, before a large group.
5. In this and the following examples, I assume that the invention has a constant, per unit return. This assumption simplifies the reasoning and, insofar as I can see, makes no substantive difference.
6. Sears, Roebuck, and Company's largest selling item.
7. For those who are not in the United States, this is famous because Sears-Roebuck lost a million dollars suit brought by the inventor, an unskilled laborer, who claimed that they had not paid him as much as they should have.
8. See my *Trials on Trial*, Columbia University Press, New York, 1982, for a general discussion of the probems of litigation and suggestions for reforms which would sharply reduce its cost. The book is addressed to the general problem of litigation and not specifically patent litigation. I will discuss Kingston's proposals for dealing with this problem later in this Chapter.
9. "Costs of Ordinary Litigation". David M Turbek, Austin Larat, William L F Felstiner, Herbert M Kritzer, and Joel B Crossman, *UCLA Law Review*, Vol 31, No. 1, October 1983, p. 72.
10. It might be sensible to provide an arrangement under which the person patenting whatever it is could, by paying extra fees large enough to cover the cost, obtain a careful examination of the sort that we now give to new patents. Providing for public notice, and then making the patent binding, even if after that point in time it is discovered that there was an article containing the same information in a Tibetan Journal which is collected in the library of the Maharaja of Nepal would be basically a minor change in the present law. The proceeding would be similar to an action to quiet title in real estate law.
11. They advertised it as a knee action.
12. And also his work as consulting engineer on his own refineries.
13. If they honor domestic patents this is, granted the small size of their market, more or less irrelevant to any economic decision.
14. Most of these small countries cannot take the easy way out, i.e., hiring several employees of the company that has the know-how and getting the know-how that way, because their true credit rating is too low. An engineer for Corning Glass is unlikely to be willing to go to Liberia even if the Liberian government offers him very good terms because he feels, quite rightly, that the Liberian government is not by any means certain to keep its promises and in any event, the Liberian government next year may be different. A stable government in a place that is pleasant to live in, like Switzerland, or for that matter the Netherlands, would avoid this problem, but so far none of the countries in this category have really made serious efforts to do so.
15. Flat glass had for a long time been available, although made by a much more expensive process.
16. For many years Japan was one of the more successful evaders of the international patent system. They even granted what amounted to local, that is within Japan, patents to Japanese citizens on things which had previously been patented elsewhere. I think one could expect a similar imaginative approach to the innovation patent or warrant.

CHAPTER VIII

Z. A. Silberston

INTRODUCTION

Kronz and Kingston argue that the direct protection of innovation would bring with it a number of advantages. Historically low (and falling) financial returns to investment in innovation would be greatly improved; employment would be increased both by a stimulus to investment and by a shift from labour-saving to output-increasing innovations; industry structure would be changed in favour of smaller firms as opposed to multinationals; direct protection of innovation would offer a dynamic means of meeting economic pressures from Japan. These are ambitious claims (although many might not think that the achievement of all of them is desirable). Even if they could not be met in full, however, there might still be a case on more modest grounds for direct protection of innovation.

In this chapter I look briefly at the proposals of Kronz and Kingston, and then examine them critically. In particular, I consider their likely practical consequences, and go on to examine the possible advantages and disadvantages of introducing legislation on these lines.

KRONZ

The aim of Kronz, and also that of Kingston, is to protect innovation directly, Kronz is a trained patent lawyer, so that his criticisms come from within the patent system, and arise from discontent with it. He wishes to protect products. Processes would not be protected directly, but would receive their protection through the physical components they contained. Imported innovations could qualify for protection. The time period of protection would vary from case to case, and this would apply to the period of any licences granted also. The system would either replace the classical patent system altogether, or supplement it, and the individual examiner of the patent system would be replaced by an expert board with a strong commercial bias in its make-up. Grants, once made, would be incontestable.

Under the Kronz system, it would be possible to obtain protection for products of too low a level of invention to qualify for traditional patent protection. Protection would be given for anything new, in the form in which it entered commercial activity. Kronz stresses the similarities with protection for copyright. Processes cannot be protected since they are not tangible objects, and these alone can be protected, but a process can be protected through its components. For this purpose, specific settings would have to be supplied for equipment which could be operated in more than one way. Anything could be protected that could be described by using a " vocabulary" of tangible objects, and which could also be commercialised. This would include printed matter and computer programs.

Innovation protection could only be given to firms, since they alone could effectively commercialise a product. Where necessary a combination of firms might jointly obtain protection. A combination of firms might also be formed to obtain a more remunerative period of protection than any one alone could obtain. Kronz conceives of protecting innovation but not diffusion, and proposes to ensure this by fixing an appropriate length of term. If the patentee withdrew, without good reason, from production of the innovation before the diffusion stage was reached, he would lose his monopoly.

A feature of the Kronz system is that the innovation need not be presented fully fledged to the Innovation Office. It could be disclosed step by step, as the process of arriving at the innovation proceeds, and the Office would take this disclosure into account when considering priority for the eventual innovation. Once commercialisation occurred, application could be made for a provisional grant of protection. At this stage the specification would be published, and third parties wishing to oppose it would have to do so at that time since no opposition would be possible after grant.

If any licences were granted, their term would be set by the Office, bearing in mind the innovative capacity of the licensee. If this were considered low, so that rewards from the licence might take some time to be reaped, the licence term might extend beyond the term of the original Innovation patent. Compulsory licences would not be needed under such a system, in Kronz's view, and there would be no provision for them.

The international aspects of the Kronz system are interesting. Importing a product produced abroad could qualify for innovation protection, just as with the present patent system. The Paris Convention would also apply, giving protection in another country within a year of protection in the first country. However, if there were no production in a country, the term of protection would be shorter than if there were. This is because not a great deal of investment might be involved in importing, and because entry into international trade can be taken as evidence of a transition between the innovation and the diffusion phase – and the latter is not given protection.

Kronz sees his system as supplementing or replacing the classical patent system. If both systems were in existence together, however, an innovation patent might infringe a classical patent, and resolution of the conflict would then be necessary. Kronz believes that in several instances the provisions of the existing patent system would enable such conflicts to be resolved.

KINGSTON

Kingston's "Innovation Warrant" resembles Kronz's Innovation Patent in many ways, but there are interesting differences. A major difference is that Kingston definitely wants to retain the existing patent system and patent office. His Innovation Warrant would be administered by a public authority which would almost certainly be separate from the existing patent office. Kronz, on the other hand, appears to envisage that patent offices would administer the new system of innovation protection as well as the existing patent system (if it remained in existence). One of the reasons for the difference between Kronz and Kingston over this issue may be that Kronz approaches these questions with his background as a patent lawyer. Kingston on the other hand sees the problems in terms of economics, and it is no doubt partly for this reason that he stresses the possible benefits of competition between the innovation office and the patent office.

Kingston is concerned with encouraging investment to produce new things. Under his scheme anything new can be protected as long as it can be the subject of investment, i.e., as long as it can be bought and sold. This means, for example, that computer software could be protected, and so could methods of doing business. The criteria would be newness and the purchase or sale of these things for money.

Stress is placed on the quality of protection. This involves the length of the monopoly period, and also the effectiveness of protection. The more perfect the protection, the shorter the warrant period could be. Kingston conceives of project-related and firm-related risk. Both would be taken into account in determining length of protection. For example, a small firm with limited resources might be given a longer monopoly period, for any given project-related risk, than a large multinational firm, because its firm-related risk would be greater.

An important feature of Kingston's scheme is his emphasis on national markets. A product to be protected would have to be made available in the ordinary course of trade for the first time, and that would mean that it would have to be available through investment for production in the national market. Imports would not satisfy the condition for the granting of a warrant, although they would for Kronz: this is an important difference between the two schemes. Foreigners would be entitled to apply for a warrant under Kingston's scheme, but they would have to make the appropriate investment in the country concerned in order to obtain a warrant there. If they did not do this, the protection given to a domestic firm could not be broken by a foreign firm exporting a similar product, even though the firm might have warrant protection in its own country. It is not clear what would happen if the foreign firm had patent protection of the existing type in both countries: presumably in that case it would be protected when it exported to the second country. The national element in Kingston's scheme would then work for warrants but not for patents.

One of the problems with the patent system is that small firms often cannot afford the expense of protecting their patents when they are infringed. Kingston would get over this by making the enforcement of the warrant the business of the authority which had granted it. He even considers the possibility that infringement might be regarded as a crime, with appropriate penalties. At the other extreme, infringement could be regarded as a tort (as with infringement of a patent). In this case the warrant holder might take legal action, but the financial burden of protecting his monopoly could be removed by generous legal aid from the Innovation Office. An alternative system would be for the Innovation Office itself to prosecute offenders in the Civil Courts. As a deterrent against infringement, exemplary damages should be available, on the model of the United States "treble damages" suit. On balance Kingston prefers the system under which the Innovation Office would itself take (civil) action.

As with Kronz, the warrant would be incontestable. There would be an opportunity for opposition before grant, but not once the warrant had been granted. The monopoly could not therefore be lost by post-grant opposition, as it can be under the patent system. The fact that a warrant would be incontestable, and protected by the State, would in Kingston's view justify very much shorter periods of protection than are available for patents.

Kingston envisages that the Innovation Office would in practice need to have no administrative discretion over the period of grant. Both project-related and firm-related risk could, in his view, be objectively determined, so that the period of protection could be easily arrived at. The applicant could, however, contest the proposed period of grant, as could its competitors if they thought it was too long.

In order to protect existing warrant holders, the Innovation Office would have to take action against competitors producing equivalent products. These would be "commercially equivalent" – a new concept. A different innovation which itself gained a warrant would not be open to attack under this procedure. However, the holder of a new warrant might have to pay a royalty to an earlier warrant holder, to compensate him for the loss of profit-earning activities that the new warrant might involve. Once again, in Kingston's view, this procedure could be operated without requiring the use of discretion by officials of the Office.

Once an applicant had been offered a warrant, he would be given a specified period in which to determine whether to undertake the necessary investment. The warrant would be granted only when the Office was satisfied that the investment was to take place, and the warrant would lapse if after a specified time the investment had not been made.

Kingston envisages that the procedure for granting warrants would act very speedily. He believes that six months from application to grant should be possible, even if there were opposition to the grant.

Like Kronz, Kingston is interested in protecting new products, not new processes. A process would be protected by what it produces, which would have to be new. However, a substantial reduction in price would count as novelty, and a process could be indirectly protected in this way.

A final feature of Kingston's proposal is that it makes provision for additional protection to be given to particular regions of the country, or to particular industries. This would be a governmental decision, in the form of Statute or Ordinance. A multiple of the monopoly term would be laid down, to apply to investment to produce new products in particular regions, or to investment to produce particular types of new products whose output the government wished to encourage. The Innovation Office would then apply the necessary multiple automatically to the term it would otherwise have granted. Kronz makes provision in his proposal for regional discrimination also, but in his case discretion appears to be given to the Innovation Office itself, rather than to the government.

COMMENT ON THE PROPOSALS

a) Criteria for Innovation Warrants and Classical Patents

The difference between the new proposal and classical patents will to some extent be obvious from the description of the innovation warrant (or patent) above. One difference lies in the degree of novelty required. The subject of the warrant has to be new, but the degree of novelty need not be great. Presumably the mere renaming of an existing product (for example, a bar of soap) would not be held to be new enough, but a slight change in its formula might be, as long as the new formula had not been marketed before. The firm putting the new bar of soap on the market would then have a monopoly for a period (although not perhaps a long period in this case), and no "commercially equivalent" product could be marketed by another firm without a licence from the first firm. Even if the monopoly term, or the conditions of any licence, were not onerous, there could be occasions when the degree of protection afforded to the first firm might be held by many to be excessive.

In these circumstances it seems to me that the criteria for "newness" might well have to be tightened up over time by the Innovation Office (and the law under which it operates).

This might apply also to products not at present within the scope of patent law. For example, the idea of harnessing the tides to make power is not at present patentable, although some devices to do this may be. If protection is given for innovation for some product which can use tidal energy, but which cannot be patented under the present system, the financial benefits might be considerable. Would public opinion really sanction large rewards for a modest degree of novelty? Already, the high returns on some patented drugs, for example, are antipathetic to public opinion. Would not the same apply *a fortiori* to innovation warrants yielding a high return?

One questions also the strength of protection offered to some "products" that would gain innovation but not patent protection: computer software, for example, or methods of doing business. It might be very easy to evade a monopoly in such things, by introducing small changes. If, on the other hand, the products arising from such changes could not get warrants because they would be judged to be commercially equivalent, public opinion might again feel that too much protection was being given for things whose degree of novelty was not great.

If I am right about the problems that might arise if strong protection were to be given to certain types of innovation, it seems to me that pressure might soon arise for the criteria for granting an innovation warrant to be made tougher. If this were done, the criteria relating to "being made available for the first time" would have to be strengthened. In that case, something like the "inventive step" of the present system might be increasingly insisted upon, and the apparent ease with which innovation warrants could be obtained would no longer apply. The innovation warrant would still differ from a patent in certain respects (for example, the need for national working), but it might not differ much in the degree of rigour required.

b) The protection of processes

Neither Kronz nor Kingston wants to protect innovation in processes, although Kronz is prepared to do so indirectly, by means of components. Kingston also is prepared to regard a substantial reduction in price as novelty, so that if a new process makes a large price reduction possible a process might be protected. One wonders whether in practice processes would effectively be protected through such provisions. The invention of float glass by Pilkingtons for example, comprised new components, and it eventually led to a big reduction in the cost of producing distortion-free glass (and its price). It is not however obvious that this process would have gained an innovation warrant. The denial of a warrant (assuming for the moment there were no patent system running in parallel) would surely have inhibited the development of this outstanding invention. Would Kronz and Kingston have wished this to occur? If so, it is difficult to understand why.

If the patent system continues in existence, as Kingston definitely wants to happen, the problem of protecting processes can be taken care of by the use of patents. In that case, the absence of the less rigorous protection of the warrant system might not be a great loss. If patents ceased to exist, however, I would regard the difficulty of protecting processes under the warrant system as a great deficiency of the proposal.

c) Assessment of length of term

Kingston spells out most clearly the method by which length of term might be determined. His notions of project-related and firm-related risk as factors which should affect length of term are interesting ones. Ideas of this sort have led to criticism of the fixed length of term under most patent systems, and to proposals for petty patents, for example. On the other hand, it seems to me that both Kronz and Kingston greatly underestimate the difficulty of arriving at an appropriate length of term.

Stress is laid by Kingston on research which should be carried out by the Innovation Office – research which would enable the Office to decide on length of term almost automatically. I am very doubtful whether research could have such convenient results. In my experience, every case is a different story. Other cases may help as a background, but the particular case under review always has unique features. I believe that, unless the Office adopted very rough rules of thumb, it would soon become bogged down in difficult calculations about length of term. Presumably negotiations between the office and firms would be involved, since firms applying for warrants would surely want to present a case for a certain length of term, and this would be bound to lead to discussions with the Office. One can easily conceive of legal representation for firms, with all the delays that this would be likely to bring.

Even if applicant firms were content to let the Office quietly determine length of term, there is no reason why a competitor should take this attitude. Opposition proceedings could presumably be brought by a competitor who objected to the proposed length of term. Since the innovation warrant would give protection for a named period of time, it would surely be right for objection to be possible on this ground, as well as on the ground of the novelty of the innovation.

The more one thinks about it, the more difficult and time-consuming the process of the determination of term for the warrant appears to be. I find it almost inconceivable that the time needed for final determination of the term could be anything like as short as Kingston claims.

d) Conflict of jurisdiction between patents and warrants

Kingston wants the patent system to continue, and Kronz probably wants this also. Kingston conceives of different offices as well as different criteria for patents and warrants. It is not clear which system would have priority if conflicts should arise.

Warrants could be given for products which would not be novel enough to gain patents, and in that case conflicts of jurisdiction would not occur, at least not on the national level. Conflicts of jurisdiction might occur if a warrant were given for a product to one firm, while another firm was given a patent for the same product. Would priority be decided by the usual rules? In that case, close liaison would be needed between the patent office and the innovation office. With two separate offices, working to different criteria, some difficult situations might well arise.

Difficulties might also arise over processes which had been patented, while products associated with new processes (and produced by another firm) might have been granted warrants. Here there would seem to be a straight conflict of jurisdiction, and the law would have to make clear where priority lay in such cases.

Another form of conflict might arise with Kingston's scheme, which requires national working for the grant of a warrant. A foreign firm exporting a product patented in the importing country might infringe a warrant granted in that country. If the warrant were given priority, then the patent protection would be useless. If the patent were given priority, the warrant would be useless. In this case also, therefore, the law would need to be clarified.

There must be many other forms of potential conflict between the warrant and the patent systems. Kingston likes the idea of competition between them. I cannot help thinking that in a situation where legal monopolies are being granted by the State, only confusion can be caused by two parallel systems. One system, if it could be devised, which combined features of both the warrant and the patent systems would give rise to many fewer difficulties.

e) The powers of the Innovation Office

The Innovation Office might have two powers in particular not available to patent offices under present legislation. The first (in the Innovation Patent) is the power to regulate licences, and the second (in the Innovation Warrant) is the power to enforce warrants, or to pay the costs of litigants to enable them to do so.

As regards licences, the situation under the patent system is that, subject to certain limitations, patentees can grant licences on any terms that are mutually acceptable. The authorities become involved only in compulsory licence proceedings. Under the Kronz innovation patent system the Office would supervise the granting of licences, and would determine the term. It is difficult to see how the Office could do this and yet ignore the royalty arrangements. In practice the Office would seem likely to get involved in the whole negotiation. This would make it the arbiter of many transactions now held to be private, and would involve much more State interference than is now the case. Perhaps this would be desirable, but one cannot help thinking that this system would inhibit enterprise and initiative rather than encourage it. In any event, the amount of work for the Office would probably be very great, and would call on substantial administrative resources.

The ability to help warrant holders financially to defend their monopolies is a new proposal in this area. It contrasts with the patent system, under which enforcement is left to private individuals and firms. The arguments underlying the proposal are cogent ones – especially the support given to small firms – but the availability of State funds may well lead to much more litigation than at present. In addition, while the patent system exists alongside, it is odd to grant funds to defend warrants and not patents, when the criteria for granting patents are likely to be much more demanding. This seems to be another area where the existence of the patent and the warrant system in parallel is likely to give rise to inequities and difficulties.

f) Small and large firms

The warrant system would encourage small firms in a number of ways. It would help them financially to defend their warrants, and it would give them longer terms than large and powerful firms. At the same time, it would protect them against imports of similar products, unless a patent took priority. Encouragement of small firms is generally thought to be desirable. It counteracts the monopoly power of large firms and provides many centres of initiative and employment in the economy. Other things equal, therefore, this tendency of the warrant system is to be welcomed.

What seems doubtful is whether these features of the warrant system would have more than a marginal effect. The financial ability to defend a patent in the courts is useful, but it is much more useful to have the financial ability to innovate and market successfully. The warrant system gives this indirectly to small firms, by strengthening their bargaining power with big firms, but it is still likely to be the big firm that has the necessary financial strength to innovate in many fields. On balance, the encouragement of small firms under the warrant system seems to me unlikely to have an appreciable effect on the pattern of rewards for innovation.

In one respect the warrant system might be unfair to really small inventors. Under the patent system the small inventor has the protection of his patent with which to negotiate with large firms. Under the warrant system, only firms can obtain warrants, and they have to show that they are able to command the funds with which to innovate. Small inventors not organised into firms (or in very small firms), and without access to funds, might be in a weak position under a system of this sort. It is true that a private inventor might be given an "option" on a warrant for several months, and would be able, during this period, to search for backing. Nevertheless, without investment he cannot proceed, and if he finds it difficult to raise funds for investment his warrant will lapse. On balance it is the medium-sized firm, rather than the small man or the small firm, that seems most likely to benefit under the new system.

g) Nationalism

One of the main features of the Kingston scheme is encouragement of production in the country where a warrant is granted. A warrant holder is protected against competition from similar imports, since an importer cannot be granted a warrant. A foreign firm may gain a warrant only by investing and producing in the country concerned. Kronz does not go quite as far as Kingston in this respect, but he would grant a shorter term of protection (for reasons given earlier) to a product that was not produced in the country. The two systems might have similar results, therefore.

While the patent system exists alongside the warrant system, as argued previously, the import of a product patented in the importing country would be protected. This domestic working provision of the warrant system might not in practice therefore be operative. Let us ignore this for the moment, however, and ask whether nationalistic protection of the sort envisaged by Kingston would be effective, and also whether it would be desirable.

Up to a point, Kingston's system would obviously protect home production. If, however production of a product were profitable in a given country, this would be likely to attract foreign firms, who might well find the country a useful base for exports as well as for sales. Attraction of this sort occurs in the present situation, where inward investment into the UK, for example, has been a notable phenomenon in post-war years. It is difficult to believe that the warrant system would add greatly to present inducements for overseas investment. It might, however, tip the balance in marginal cases, when a foreign firm is wondering whether or not to invest in a country.

From another point of view, protection of the home market could have deleterious effects on a country's welfare. Production at home could be relatively inefficient, while production overseas might be efficient and low cost. The consumer (as opposed to the home producer) might benefit greatly from low-cost imports. Is the encouragement of inefficient home industry to be given priority over the welfare of the consumer? It is not evident why this should be so, unless one assumes that any jobs sacrificed at home in a particular industry will never be replaced by jobs in other industries and services. Such an outcome might seem likely when the economy is depressed, although regional schemes might be of some help (and it should be remembered that Kingston himself has a scheme which might help regional employment). In any event, additional employment elsewhere in the economy is likely to occur in due course. The national economy would benefit from a change in the structure of the economy in the direction of greater comparative advantage.

The nationalistic features of the warrant scheme are subject to the same objections, therefore, as protection in general. This applies to Kingston's desire to protect the UK and other countries from the Japanese. It is by no means clear that, in the long run, countries competing with Japan actually sacrifice welfare as a result of Japanese scale and efficiency. On balance, it seems to me that they are very likely to benefit, although they are of course likely to be subject to severe transitional problems.

It is true however that the nationalistic element in Kingston's scheme is limited in duration, because the term of the monopoly is limited. The protection given is also limited in scope, compared with other forms of protection, since it only covers innovations. It is much shorter than is likely to be the case with other forms, such as tariffs or quotas, which nearly always last too long. It is true also that all countries attempt to give assistance to innovation which is additional to that provided by the patent system. In this respect there is nothing unusual in Kingston's proposal. What is unusual about it, however, is that it brings a much more nationalistic flavour to the protection of intellectual property than has hitherto been the case. In Kingston's eyes this is an advantage, of course, but I have my reservations, as has been seen. In any event, the parallel existence of the patent system would greatly complicate the degree of protection actually available under the Kingston system.

g) Conclusion

Kronz and Kingston have a number of aims in mind in proposing their new form of protection. They want to plug loopholes in the patent system, to introduce more flexible provisions for length of term, and also to support small firms and domestic production. Kingston, in particular, is especially concerned with the threat of competition from Japan, and its adverse consequences for Western industry.

I have expressed some scepticism about the extent to which the innovation warrant could achieve any of these aims. At the narrow technical level, I foresee snags about deciding on length of term, for example, which might obviate one of the claimed advantages (the simplicity) of the new procedure. I also foresee grave problems if the patent system co-exists with the warrant system. At a broader level, I doubt how far an innovation of this sort, in the intellectual property sphere, can appreciably modify the forces exerted by the big battalions ("capability", in Kingston's words), although I do not deny that it might have some impact.

The proposed co-existence of the classical patent system, on the one hand, with the Innovation Patent or Warrant (or some amalgam of the two), on the other, seems to me to be a great obstacle to these proposals. I foresee serious problems arising from this. At the same time, what seems to be the logical proposal – the wholesale replacement of the patent system by the direct protection of innovation, as proposed – would be an administrative and legal task of enormous magnitude. It would take years to obtain agreement on it and years to carry it out. It would be very hard to justify attempting such a reform without very strong arguments for the changes advocated, and I question whether such strong arguments exist.

Given these conceptual and practical problems, it seems to me that the direction in which Kronz and Kingston should move is that of attempting to modify the patent (and copyright) system. This system has been the object of much legal and administrative development in recent years (with the foundation of the European patent, for example), and its supporters are going to be very reluctant indeed to throw it overboard. I can understand the impatience felt by Kronz and Kingston with the patent system, but I am afraid that I remain sceptical as to whether their new proposals can simply be added to the existing system. The introduction into the patent system of a wider adoption of petty patents, for example, would not satisfy Kronz and Kingston, but nevertheless this kind of modification might in practice be the most likely to occur in the foreseeable future. In any event, radical change along the lines they wish to see would be anything but easy to achieve.

CHAPTER IX

Henk Wouter de Jong

INTRODUCTION

The proposals for "Direct Protection of Innovation" in Chapters II and III, as well as the accompanying explanatory text in Chapters I and IV, do seem to me to be both original and of great importance. During recent decades, economists have debated extensively the value of the existing Patent System and many contributions to this debate have voiced criticisms of smaller and larger importance. Some have even advocated abolishing the system altogether. The arguments in Chapter I of the Report are in line with these criticisms. What is particularly valuable in this Chapter is the emphasis laid on the change which occurred in the principle of patenting, leading towards the indirect protection of innovation instead of direct protection. Many consequences have issued from this change, most of them with a socially negative impact. This is the reason that the three authors of a famous book in the field could write : "It is almost impossible to conceive of any existing social institution so faulty in so many ways. It survives only because there seems to be nothing better".(1) The alternatives put forward by various authors have been extensively discussed in the literature, where they were found to contain many drawbacks. Some proposals of this type related to Government awards for inventors individually or in a centralized fashion (as occurs for example in the T.N.O. system of the Dutch Technical and Scientific Organisation, which has to promote the application of scientific research in society and to advise the Government in these matters). Other ideas, related to compulsory licensing, to schemes for assisting inventors, to reductions of taxation and so forth, were all found to have disadvantages as well.(2) The question now before us, is therefore whether the proposed alternatives – of protecting innovation directly, instead of indirectly – will fare better. The two most crucial questions in this respect are, first, whether the proposed alternative is feasible and, second, whether the claimed advantages are real and/or whether the alternative system would not have other disadvantages.

DIRECT PROTECTION OF INNOVATION

As was discussed in the introduction, the supplementing of the present patent system by direct protection is desirable – provided workable arrangements are feasible and the costs are not too high.

It seems to me that the best way to comment on the proposed alternatives of direct protection of innovation is to make remarks on what is said at the beginning of Chapter IV. There, the points of agreement between the Kronz and the Kingston proposals are set forth, as well as their four main differences. A discussion of both aspects should make clear the merits and demerits of the intended changes. I will take up the main points of agreement and disagreement as they are enumerated on the pages cited and draw up the balance between them in a conclusion.

(a) *The subject matter of protection*

The proposal of both Kronz and Kingston is to protect innovation instead of invention. But what is an innovation? It is said to be "anything which can be embodied in marketable new things" (as in the Kronz proposal) or investment that is concerned with getting new things done (as in the Kingston proposal). In both cases the innovation is not restricted to new technology or technologically new products, but covers other things as well as long as they can be sold in the market. These seem to be clear criteria, but in my opinion some problems do not appear to be sufficiently discussed. First, as in the old Patent System, it has to be decided what is "new" in the products or investments for which patent protection is sought. It seems to me that to establish whether a product or a process is new is at least as difficult as to verify whether an idea is new. As will be discussed below, economic literature has not been able to provide us with a workable definition of an innovation. The protection has also to be applied to a much wider field because not only technological goods and processes are covered but also goods with a lower level of invention could find protection, as well as non-technical innovations. This extension of the area of application is somewhat reduced because the new system would only have national or regional coverage. However, the net effect would probably be a widening of the task of the granting Authority, in comparison with a Patent office, the more so as many more people would feel involved by the granting of monopoly positions in the fast-expanding commercial services sector.

Second, it will also not be easy to establish what are "marketable" new things or whether "investment" is involved in getting new things done. The simple offer to sell by means of an announcement or catalogue does not make a product marketable and could be used as a pretext to block others from introducing similar products. The first two points are linked in so far as a number of new products (e.g. fashions, styles, services) are dependent for their success in the market upon others doing the same. If a car maker could get a monopoly for some new fashion (e.g. fast-backs, covered wheels, etc.) for which a patent is now not obtainable it might well be that the market would not develop at all or, conversely, that the innovating firm would be given an undeserved monopoly for something very trifling.

Third, as is pointed out in the Report, in the granting of protection to innovations, the concept of innovation would have to be delimited and distinguished from the stages which precede the first commercial act, and from those which follow during the diffusion of the product or process. In other words, invention, innovation and diffusion would have to be delimited by the granting Authority in a great many cases. This is conceptually and empirically a difficult and arbitrary task for whatever Commission, however composed (compare the difficulties competition policy authorities have with market delimitations). This task is not solved by reverting to arbitrary rules such as Kronz suggests, in defining the innovative phase as the time necessary to satisfy half the total demand for the innovation object. Nobody at the commercial start of new products, whether ballpoints, fast-back cars or personal computers was or is able to tell the total demand to be expected. In Commissions or Authorities this effort to predict the future would only produce a cacaphony of opinions, based on various guesses. These three criticisms relating to the concept of innovation (what is new, what is marketable and how does one delimit an innovation) follow from the basic fact that an innovation is only distinguishable *ex-post*. No operational definition of an innovation can be given in order to decide *ex ante* that a product or an act by economic subjects falls in this category and not in the category of normal products, processes or acts. Surveys of the economic literature about innovation make it clear that the abstract definitions as given by the authors lack practical, operational value, which is why authors choose proxies. Schumpeter, for example, uses a proxy for his general definition, the setting up of a new production function, or a new combination of production factors – the founding of a new firm, the operation of new plant and the activity of new men. Schmookler, who talks more about inventions than about innovations, nevertheless tries to cover the whole field and distinguishes between inventions ("So new that it is not known") and sub-inventions ("an obvious change in product or process").(3) But the operational difference is the possibility to patent the invention and not the sub-invention. For our purposes, this leads to circular reasoning. It may also be the case, as Scherer has argued, that the triplet invention, innovation, diffusion is not the best explanatory scheme. Scherer says: "The trouble with this schema is that it leaves in an ambiguous state the costly technical activities which are the heart of modern research and development programs. It seems more useful to describe the pre-imitation or innovative process in terms of four essential functions: invention, entrepreneurship, investment and development".(4) And he goes on to argue a point of view shared by many reseachers in the field, that during this sequence of events, the uncertainties are reduced by an order of magnitude, while the outlays increase by such an order "before an innovation is brought to the point of commercial utility". If this point of view is true, the decision an Authority has to make – what is an innovation? – becomes very difficult, and, moreover, the use of the operational criterion "first commercial use" might well favour the big corporations. In these respects the Innovation Warrant would be better than the Innovation Patent by shifting the decisive criterion one stage earlier in the sequence as depicted by Sherer. An additional – and no less important – advantage of the Innovation Warrant would be that a market for Warrants could develop where innovating entrepreneurs, who would pass through the initial stages, but not the later (development) stages would be able to trade their rights with corporations. This also eliminates the need for a "substitute innovator" as is contained in the Kronz proposal. It seems to me that this "substitute innovator" is an impossible figure for the reason that the negotiations, which have to precede the conclusion of a contract between the innovating firm or person and the substitute innovator, are about something which is as yet unprotected.

The innovator has to disclose pertinent information about his innovation so that the substitute firm will be enabled to determine its interest and evaluate its risks. If an Innovation Patent is only given to a useful, new commercial product or process it is paradoxical to require negotiations between firms before the granting of a monopoly, while the granting of the monopoly is dependent upon novelty. As the Innovation Warrant rightly expresses, one can only negotiate a right which is legal, not an unprotected idea, model or prototype.

(b) *What to protect: Technology or also other objects?*

It was already pointed out that at first sight it seems economically sensible not just to protect new technological goods and processes, but also new economic goods relating to other fields. However, this would probably increase the burdens of a granting Authority in no small way, because the field

of fashion, designs, organizations, sales methods, services (computer programmes, information systems, management practices, etc.) and the like is
very wide. Applications could be very numerous, and their contents complex.
And it is highly questionable on second thoughts whether society would
benefit from the giving of monopolies for such types of innovations. Take two
examples: the divisional structure of business firms and the supermarket. In
the first case, this innovation was initially practised by large U.S. corporations,
in order to meet their problems of diversification and growth.(5) But all
growing firms at some time in their development need this organisational
device and a monopoly could thus seriously block their natural advance, the
more so as several variations could easily be thought up and patented by some
organisation with initial experience. The supermarket concept came up when
the motor car made larger scale shopping a possibility, labour shortage cut
shop services and consumer incomes had risen sufficiently. Thus, the innovation – very simple in idea and execution – was heavily conditioned by societal
circumstances. Protecting this innovation by giving a monopoly to a particular
firm, would have seriously delayed desirable economic change. It would have
done on a national scale what is sometimes done on a local (community) scale,
when community authorities refuse permits to outside competitors because
they threaten the established retailers who have vested interests.

The difference between a technological innovation and one in methods,
organisation, services and the like is, that the second type of innovations have
much more the character of being applicable to total companies and even
whole branches of industry or in general, (6) than technological innovations,
which are specifc and less general. Thus the damage from monopoly is
commensurately much wider. Also, the risk of longer duration of the monopoly is higher, because these aspects of business depend upon experience
instead of research. This drawback might be lessened by a reduction of the
period of protection given to these types of innovations. For these three
reasons – higher burdens upon the granting Authority, wider monopolies and
protection which favours experience and time spent in business – I would not
be in favour of extending the area of protection too widely beyond technological innovation. The economic history of the Western world shows that the
non-technical type of innovations have sufficiently come forward when the
economic circumstances required it and were copied quickly.

(c) *The variable term of monopoly grant*

Because the two proposals differ substantially with respect to the variable
term of the monopoly grant, I will discuss them in turn. In general it can be

said, that whereas a variable term certainly has the merit of being more fair and efficient, it is inevitably more complicated than a fixed term of protection. All depends therefore on the ways in which variable terms can be applied.

The *Kronz* proposals are individualistic in nature in that they make the length of the term dependent on both the "innovative capacity" of the applicant and the likely evolution of the market. An important feature is also that there is no provision for the extension of an awarded Innovation Patent. The system proposed makes it necessary for the Innovation Patent Office to be right about the term immediately it grants a monopoly. For too short a term granted might easily lead to failure to carry out the innovation, for want of sufficient protection; too long a term granted would burden society with undue monopoly positions.

But it seems to me that no Patent Office whatever will be able to settle the right term by assessing innovative capacities of firms and the likely evolution of a market. Both assessments are extremely difficult to make. Not only because much has to be known about the firms and markets concerned, but also because comparisons between firms from different branches of industry and from different segments of the same branch are implicitly required. Otherwise, the granting of terms would become a chaotic and purely arbitrary game to be played by the Patent Office. Is Philips a firm with more innovative capacity than Siemens? What *is* "innovative capacity"? Financial resources, managerial drive, quality of researchers, experience in the field, capability to organise missing information, etc. or some mixture of these qualities? And is the required mixture a variable linked to the type of industry, the product or process, its stage of development and similar features? Is Philips a firm with a better innovative capacity than Unilever, Hoffmann la Roche or some important aircraft producer? It seems to me to be a sheer impossibility to cultivate objective standards in this respect and no person on earth can get further than voicing "opinions". It is clearly impossible for a Patent Office – which is composed of assessing *individuals* – to tackle such rating tasks adequately.

The same applies to the second factor: the likely evolution of the market. Anyone with a minimum of experience in the researching of markets for new products or processes would raise his hands in despair when asked to estimate the future evolution of such a market, in particular the quantitative dimension. We know sufficiently from the past to say that even insiders in a trade can diverge wildly: years ago a salesman for a Dutch multinational firm estimated the European market for a new U.S. container, made of steel and plastics, at 120,000 p.a. within five years. My market research investigations gave an estimate of between 25,000 and 30,000 whereas the manager of the Hamburg plant said: at best 15,000, but probably less. The firm has never been able to sell more than 12,000 items of this type! Similarly, estimates for motor-cars in the fifties and sixties (there have been at least eleven of them by reputed institutes and authorities) were widely apart, and – with hindsight – mostly wrong. For computers the same story holds (7) as well as for other products. It is an illusion to think that a Patent Office would be able to make substantially correct assessments of market evolution. Because of this incapability to assess both factors adequately, the rest of the Kronz proposals – licensing, co-property, interruption of innovation phase, option claims, anticipation, etc. – fall down. Clearly, the Kronz proposal saddles the Patent Office with an impossible task. On top, there could well be big differences in practice between national offices and policy aspirations between countries might easily creep in.

The Kingston proposal, on the other hand, offers a number of features which make it more workable. Among these are the setting up of an Innovation Office alongside the present Patent Offices, which makes introduction and working of the new system easier; the enforcement of the granted monopoly by the Innovation Office, which would be a very great improvement in comparison with the present system; and the much simpler approach to the variable term. The latter boils down – as the schema in Chapter III indicates – to three or four terms, of roughly 4, 7, 11 and 16 or 17 years, depending on the type of innovation. In contrast to the proposal, I would prefer to simplify the intended system still further by eliminating the so-called firm – related risk. In that case only three or four categories would remain, viz. incremental innovations, technology–transfer – type innovations, important (risky) innovations, and radical (or breakthrough) ones. The reasons for this further simplification are the following: First, it could well be argued that it is wrong for the state to differentiate the treatment of its subjects by distinguishing between firm-related risks. After all, it is none of the business of a state to separate independent firms into categories and give them arbitrary treatment. The giving of monopoly positions should be related to objective criteria, which apply to all subjects in an equal way. If it is argued that "equal treatment" is only provided by the state taking into account a firm's resources which can be exposed to risk-taking, my answer would be that the state assumes a responsibility which is not hers. Economic society is not a school-class where the state has to grade its subjects according to innovation potential.

Second, this argument is reinforced by the consideration that an Innovation Office could be hard pressed to make correct assessments about firm-related risks. The level of risk – it is said (Ch. III) – depends upon how serious for a firm total failure could be, and this in turn can be held to be a factor of the size of the investment in relation to the resources available to the firm. Even if we accept the definition (and a number of critical remarks can be made here: are there not other definitions, what is "total" failure, is not partial failure sometimes sufficient to set the firm back years?) it does not follow that the operational measure would be acceptable. Concepts like "size of the investment" and "resources available" are very difficult to establish objectively and no authority is qualified for the task. Neither is an authority able to value the relationship between these two magnitudes and to link quantitative indicators of the relationship with "levels of risk". The latter is also dependent upon how managements handle their resources, the more so as the "size of the investment" is at best only an initially – given magnitude and may change in the course of time. Two firms may have the same resources and devote an equal amount of investment to an innovation and yet come out with different policies and results.(7) The Innovation Office cannot adequately, objectively and impartially value these givens, even if it could lay its hands on all relevant facts. Even the latter has to be doubted.

Third, the proposal to make enforcement of a monopoly the task of the Innovation Office, to which I subscribe, would take away many of the risks related to innovations, particularly those of small firms. What is left is an entrepreneurial risk which the state cannot and should not take over.

In sum, a categorisation of innovations into three groups related to project-risks would already be an important step forward from the present system and would introduce the possibility of varying the protection given in terms of years. Even that step would already be difficult enough because all valuations are of necessity subjective and arbitrary. Here the law holds true that the more differentiation (in terms of years) is brought into the system the higher becomes the subjective element in valuation (and the reverse). In contrast to what is said in Chapter III, the increase in sophistication of the measurement would not be an advantage, but would make the system less workable and more arbitrary. Empirical investigations cannot be expected to improve the statistical assessment of risks and the perception thereof, because innovation risks are *per definitionem* unique and highly different between products, branches and countries. As was evidenced before the E.E.C. Commission in the cigarette case (Commission vs. Rothmans International/Philip Morris) advertising campaigns on behalf of existing – and new – brands were widely different in their effect as between countries and products. In advertising there is no statistically ascertained relationship between money spent and sales outcome. This also applies to innovation projects.

Fourth, it is not true, as is claimed, that the warrant system eliminates administrative discretion, because all categorisation is discretionary and the allotment of a particular project to a category is always so. The staff making the decisions (they are compared with judges!) would have to have guidelines and again, these introduce discretion. Later on, Kingston increases this element of discretion greatly by introducing the possibility of changing the terms granted in accordance with rate of innovatory change in society or science, regional or other types of policy.

What is seen in the report as an advantage of the variation of the terms granted, is in my eyes a great disadvantage, because it introduces all the possibilities for the making of errors and the misuse of power for which states are known. The lessons from so-called industrial policies by Western states carried out during the seventies have clearly highlighted the abuses which were – unintendedly as well as deliberately – brought about by state policies.

If this behaviour is extended to innovations by means of variable terms, the public policy mistakes would be extended. In general the state or a state authority, such as an Innovation Office, is not qualified to assess entrepreneurial risks, as was brought forward by a survey of the experiences of state investment companies in a series of European countries.(8) It follows that the simpler and the less discretionary an innovation protection system is, the better it will function. In this context the recipe should be to make the variation of terms also, simple and non-discretionary. In practice it means that I prefer a three-category system, invariably applied to all innovations.

(d) *Third party involvement*

That examination under the newly proposed system relies heavily upon third party involvement is logical because not only new ideas but also new practices are taken into consideration in giving monopoly grants. As such, it would seem to me to be a step forward if society and individuals are more actively concerned with the process of granting of patents. It could well have the effect of making society more conscious of the importance of innovation for daily practices in business and also raise many people's capacity to distinguish between the positive and negative aspects of monopoly. On the other hand, third party involvement also increases the burdens of the Innovation Office, because administrative procedures will be more complicated, hearings have to be held and the Office has to be alert to the possibilities of false counterclaims, delaying tactics and the like. As these would not just relate to new ideas, but also to new practices, they might well involve more investigative tasks.

(e) *Grants and renewal fees*

That grants of monopoly should be incontestable, except in so far as they were awarded on the basis of fraudulent information to the Office, would seem to be an unqualified advantage. However, in conjunction with what was said under (d) this advantage requires for its application a more than normal caution in granting the monopoly, which might lengthen the time of investigation. Errors and faulty judgements on the side of the Innovation Office need to be avoided at practically all costs, which again would seem to be more difficult in the case of new practices than in the case of new ideas.

The abolition of renewal fees is both logical and sensible. Presently this is only a source of discrimination between firms which have the money to pay and those which have not or have it to a much lesser extent.

(f) *Independent Authority*

That the new system has to be applied by an independent authority is, after what has been said before, an unavoidable necessity. This requirement implies at the same time that the policy considerations, discussed in the context of the variable term of monopolies granted, have to be ruled out. One cannot at the same time, have an independent authority, which grants monopoly positions to particular firms and which also takes into account policy directives from Governments, aiming at varying the terms of monopoly grants in accordance with desires to temper or accelerate the rate of innovation, regional differences, etc. Like existing Patent Offices, an Innovation Office would have to be independent *vis à vis* Governments and other policy makers.

The question about the internal structure of an Innovation Office will not be taken up in this comment, because this would obviously be dependent upon the tasks which the office would acquire. The divisions which Kingston outlines in Chapter III seem *a priori* sensible, though I would prefer to define their operations somewhat differently, in accordance with the simplified tasks of the office outlined above. But clearly, research and enforcement would become as important a part of the working of the Office as would be the examination procedures and the screening for novelty.

Conclusion

It has been argued in this comment on the proposals for direct protection of innovation, that the present patent system is in need of fundamental change or even total replacement if a better alternative could be found. Many experts agree to a greater or lesser extent with the criticisms voiced. The proposals for direct protection of innovation are original and their principle as such can be accepted. It is indeed better to give monopoly protection to an existing object or process that is new, than to the idea – worked out on paper – which might underwrite a diversity of economic objects. For ideas can very often be embodied in various ways, can be circumvented, can be litigated and can be pooled by means of cross-licenses, supporting market power. On the other hand, direct protection of innovation guarantees against imitation, and if enforcement against infringement is taken out of the hands of the owners of the monopoly, a real benefit from the innovation would be secured. This might be either through self-exploitation or by means of selling the rights (warrants) concerned.

The real problem is whether the proposals are feasible. In this respect the Innovation Warrant has better prospects than the Innovation Patent, because the latter requires a judgement of the "marketability" of the innovation and of the "innovative capacity" of the firm. Both are undesirable and, above all, impossible to carry out objectively by the granting authority. A general criticism is that innovations are difficult to define operationally *ex-ante*; they can mostly only be recognised *ex-post*. The delimitation of innovations would, I fear, be the source of many difficulties. In any case, the tasks of a granting authority would be much larger than those of a present patent office.

Both proposals opt for variable terms. This seems to me unfortunate, because their introduction would greatly complicate the operation of the system, could only be done arbitrarily and might introduce unwanted policy considerations. Much better would be a system operating with a small number of fixed terms, adapted to the importance of the innovation. These terms should only be project-related, not firm-related. The latter is discriminatory before the law, is none of the State's business, cannot but be highly arbitrary and would – because of errors – discredit the Innovation Office. Several commendable features, such as incontestable grants and public enforcement complete the basic idea, which it is worthwhile to investigate further.

NOTES

1. J Jewkes, Sawers and R Stillerman, The Sources of Invention. Macmillan & Co. Ltd. London 1961. (Reprint of the 1958 edition), p. 253.
2. See the discussion in Jewkes, Sawers and Stillerman, op.cit., pp.253-260 and F M Scherer, Industrial Market Structure and Economic Performance, Rand McNally & Company, Chicago, 1980, Chapter 16.

3. J Schmookler, Invention and Economic Growth, Harvard 1966.
4. Scherer, op.cit. p 350.
5. A D Chandler, Strategy and Structure, Cambridge (Mass.) 1962, p. 78–113.
6. The multidivisional structure of a firm was quickly adopted by the top U.S. firms in various branches. See Chandler, The Visible Hand. Cambridge (Mass.) 1977, p. 474–476.
7. See Forrester, The micro-electronic revolution, Oxford 1980, p.19.
8. See Brian Hindley (ed.), State investment companies in Western Europe, Picking winners or backing losers? London 1983, in particular his own introductory article. Surveyed were Italy, The Netherlands, Belgium, The United Kingdom, Sweden, France and W Germany.

CHAPTER X

Brian D. Wright

ON THE DESIGN OF A SYSTEM TO IMPROVE THE PRODUCTION OF INNOVATIONS

The concept of patent protection of inventions has had great theoretical attractions for economists. It enables the holder to capture returns to new discoveries which could not be otherwise obtained by private transactions. Thus a private incentive is created which encourages firms and individuals to devote resources to inventive endeavors. In organizing their efforts they have an incentive to obtain and apply information regarding the nature and opportunity cost of their own relevant capacities, the cost of other research inputs, the duration of the project, the probability of success, and the value of a successful outcome. Some if not all of these attributes of a research project are quite uncertain *ex ante*; in general, the information available to interested private parties is far superior to that available to public authorities. In this situation the advantages of decentralization afforded by a perfect patent system seem unusually clear.

But the current system, as William Kingston (1985) argues convincingly (Chapter I), is far from perfect. It has drifted away from the original conception of direct protection of "new manufacture" to indirect protection via the rights to a novelty which could be used as an input into new manufacture, in exchange for "teaching" that novelty via formal disclosure. Furthermore the extent of protection offered for those without large financial resources is reduced by the heavy reliance on *ex post* legal challenge as the means of determining patent validity. The high costs of such legal actions mean that for small firms and individuals the award of a patent has very little value, while for those holding large financial resources even patents of dubious validity may be very strongly protective. Furthermore any requirements that patents be utilized if protection is to be maintained have become largely ineffectual. Powerful firms can therefore use patents to block, rather than to encourage, technical advance.

Another key argument against the current system is that the criterion of non-obviousness has been too strictly applied. The result, it is claimed, is that many worthwhile innovations which would be considered obvious *ex post*, but are not obvious *ex ante*, receive no protection and remain unperformed.

Thus current patent systems leave much to be desired. As outlined by Kingston (Chapters II and III), Kronz and Kingston have independently proposed direct protection of innovation through an independent authority. As summarized in Chapter IV, some of the common features of the two proposals are that any saleable object not previously or currently "available in the ordinary course of trade" qualifies, that the monopoly grant has a variable term, that third parties are heavily involved in approval *ex ante*, that the grant is contestable *ex post* only on grounds of fraud, and that the terms of grant may be specific to a region or country.

There are four very important differences between the proposed Innovation Patent and Innovation Warrant schemes. First, the length of term of the Innovation Patent is adjusted to match "innovative capacity" of the Patentee in each case, whereas the term of an Innovation Warrant is chosen from a prespecified set (in the example, six terms ranging from 4 to 16.7 years). In both cases, however, firms with lower innovative capacity receive longer terms on their patents, and higher risk means more protection. Innovative capacity includes factors collectively described as "capability," but in Kingston's proposal the measure is simply size of investment relative to net assets of the firm.

A second difference between the two schemes is that the Innovation Patent scheme requires that the item to be patented exist before protection is given, whereas the Innovation Warrant scheme stipulates conditions for progress on the innovation after the initial award. Third, the Innovation Warrant is to be protected by and at the expense of the State, while the patentee has this responsibility in the case of the Innovation Patent, as in the present system. Fourth, the Innovation Patent would replace current invention patent schemes, whereas the Innovation Warrant is offered as a complement to current arrangements.

Both of these kindred proposals show the benefits of years of experience, observation, and careful reflection upon the current flawed systems of protection of invention. They have much to commend them, and I should be surprised if they do not influence the course of future legislation. However my allotted task here is neither to praise nor to bury the proposals but to act as a constructive critic. In what follows I make no attempt at a balanced review, but concentrate on areas where I believe special attention is warranted.

Like much of the work in this area, the proposals emphasize the incentive problems caused by lack of appropriability. But very little attention is directed to optimizing the supply of innovative effort. Supply considerations affect the choice between the *ex ante* award of Kingston and the *ex post* award of Kronz, and also, unfortunately, greatly complicate the identification of the optimal variable term.

Of course, if supply is essentially unresponsive to incentives, its neglect is appropriate. But then incentives are also irrelevant, except as mechanisms of income distribution. Important works on invention patents, including those of Plant (1934) and Machlup (1958), have taken essentially this view on the grounds that invention is the work of persons with peculiar gifts whose efforts are motivated by factors largely independent of monetary reward. Whatever one's view of this position with respect to inventors, it is much less justifiable with respect to innovators. Innovation is essentially the act of embodying inventions and other new ideas or constructs in actual products. As Kingston (Chapters I through IV) emphasizes, this process depends largely on the willingness of investors to make funds available under conditions of risk. The supply of innovation, in other words, is much more like the supply of other goods produced by factors available at approximately constant cost over the relevant range of output. In economic terms, we should expect, at a given incentive level, that the supply of innovation will tend to be highly elastic.

In what follows, I first show how supply conditions affect the optimal choice of term of the award. Then, in Section 2, I consider the question of timing of the choice of recipient of the award, that is, whether it should be decided prior to the achievement of the innovation, as in the Kingston proposal, or after the innovation has been achieved, as in the Kronz plan. I present some observations on aspects of the two proposals designed to match the term of the patent to the characteristics of the prospective innovator in Section 3. Conclusions follow in Section 4.

1. *Optimal Choice of Term When All Innovators are Identical*

The term of patents (except petty patents) in most countries is between fourteen and twenty years. The wisdom of choosing a term in this range is broadly supported by a traditional analysis, formalized by Nordhaus (1969) and further exposited by Scherer (1972). This analysis interprets the patent life limitation as a tradeoff between the welfare cost of the restriction of use inherent in the exploitation of the patent right and the incentive to invent afforded by that right when the benefits of invention cannot otherwise be appropriated by the inventor. The conclusions are that the term should be at

230

least as long as the traditional two terms of apprenticeship (fourteen years), but beyond that length, the social value is quite insensitive to variations in length and to differences in demand for and supply of invention. Hence the exact choice of patent life in a given situation is not very crucial, and adjustments for different types of invention are unnecessary for supporters of current patent systems.

The analysis I shall present below leads to quite different conclusions, much less favorable to the status quo. But to understand this alternative argument, it will be helpful to consider the essentials of the Nordhaus-type analysis using a very simple model chosen purely for its heuristic value.

1.a. *The Traditional Analysis*

Consider the single-period example in Figure 1, which illustrates the market for a competitively produced consumer product, Q, subject to a cost-reducing invention. The demand curve is D and the marginal revenue curve is MR. In the initial situation the supply curve (the marginal cost curve of Q) is S_0, and production is Q_0. If an innovation that reduces the marginal cost by $P_0 - P_1$ is made freely available, then the new supply curve is S_1, socially optimal output

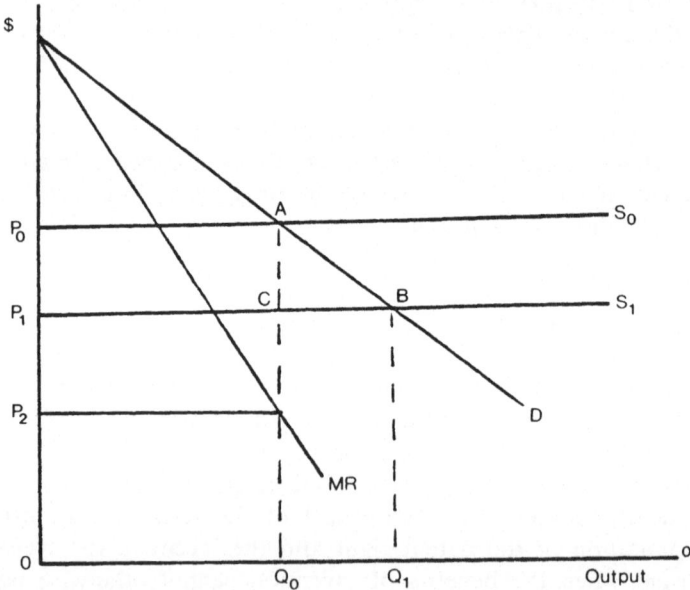

Figure 1. The Welfare Cost of a "Run-of-the-Mill" Patent Caused by Restriction of Use.

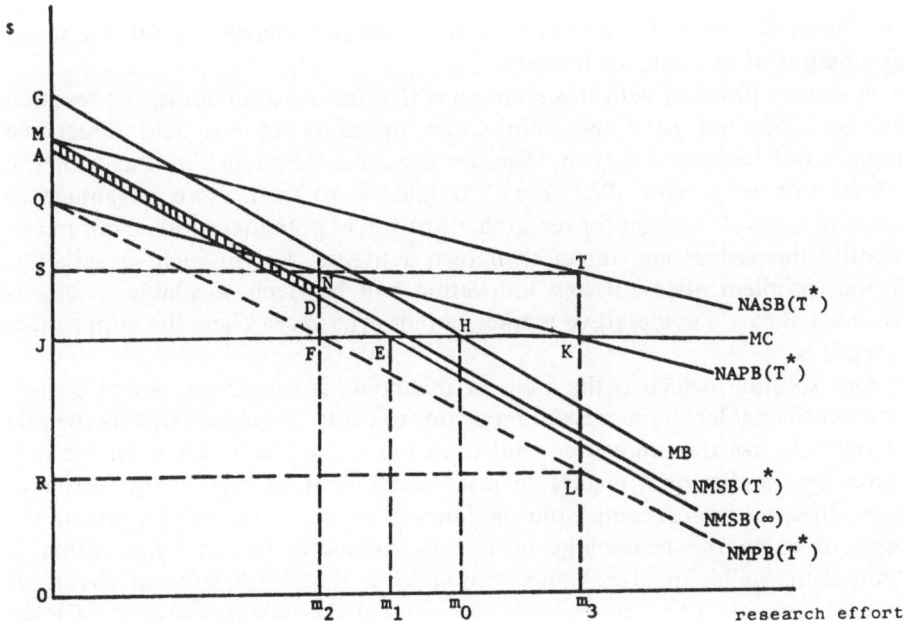

Figure 2. Determination of Optimal Patent Life: The Conventional Analysis Reconsidered.

is Q_1, and the social value of the innovation, B, is the increase in consumer surplus, the area P_0ABP_1. (If supply of Q were less than perfectly elastic, one would have to be careful in specifying the type of supply shift (e.g. parallel or proportional) induced by the cost-reducing innovation. Some of the social benefits would then appear as increases in rents to producers of the good Q rather than as consumer surplus.)

For simplicity assume that this one, cost-reducing innovation is the sole objective of researchers in this area, and that researchers are competitive and risk-neutral. The probability that at least one of the researchers will find the innovation is $\mu(m)$, where m is the aggregate research effort supplied. Then the expected social value of the research is B. $\mu(m)$, and the marginal expected social value of research is MB $\equiv$ B. $\mu'(m)$, where a prime (') denotes the first derivative of μ with respect to the argument, m.

If, as one might expect, the marginal probability of success is positive but decreases with m (the second derivative, $\mu''(m)$, is negative) then a risk-neutral social planner would maximize expected net social benefit at the level of m at which expected marginal benefits MB equal marginal cost of research MC. In Figure 2, where MB (and, by implication, $\mu'(m)$) is, for simplicity, assumed to be a linear function of m and unit research costs are constant at OJ, the social optimum is at m_0 units of research.

A serious problem with this scenario is that in many situations the research planner does not have the information necessary to hire and direct the suppliers of research effort m. (See, for example, the model with asymmetric information in Wright 1983, 1985.) If that is so, there is an advantage to opening a private market for research effort where potential suppliers of m can identify themselves and direct their own activities. But in such a market, a further problem arises: if any innovation will be freely available to all, as assumed above, a competitive market cannot offer researchers the appropriate incentives.

One solution, which is the focus of this book, is to provide patent protection which enables the successful innovator to exploit monopolistically the sale of rights to use the innovation. But if an innovator has a patent on the new technology and cannot engage in price discrimination (i.e., charge different users different prices) rents from the innovation are maximized by setting the price of using the technology at $P_0 - P_1$, assuming that the innovation is "run-of-the-mill," in Nordhaus's terminology, that is, $P_1 > P_2$ in Figure 1. Output remains at Q_0. In this case the value of the patent is area $P_0 A C P_1$ in Figure 1, and it all accrues to the holder of the patent. Compared to the free-use case, the surplus generated has been reduced by the welfare cost, L, represented by triangle ABC. Given the patent award, in Figure 2 marginal social benefit is $NMSB(\infty) \equiv (MB - ML)$ where $ML \equiv L$. $\mu'(m)$; this equals marginal cost at a level of research of m_1 units, and is interpreted by Nordhaus and others as the competitive equilibrium with infinite patent life.

How does a limit on patent life affect this analysis? Assume that annual returns would have been constant forever if the patent life were infinite and that the interest rate is constant at r. Then if the patent life is limited to T years, the present values of profits and of excess burden are both reduced by a factor $\alpha(T)$,

$$\alpha(T) = 1 - \int_0^T e^{-rt} \, dt.$$

Given $\alpha(t)$, the present value of the marginal private returns is $(1 - (T))(MB - ML)$ and the net marginal social return is $NMSB(T) \equiv (MB - (1 - (T))ML)$. As the patent life, T, decreases from infinity, the marginal private return at any level of M obviously decreases. But the marginal social value rises toward MB, the marginal social benefit at $T = O$, because the welfare loss from the monopolistic restriction on use of the patent occurs over a

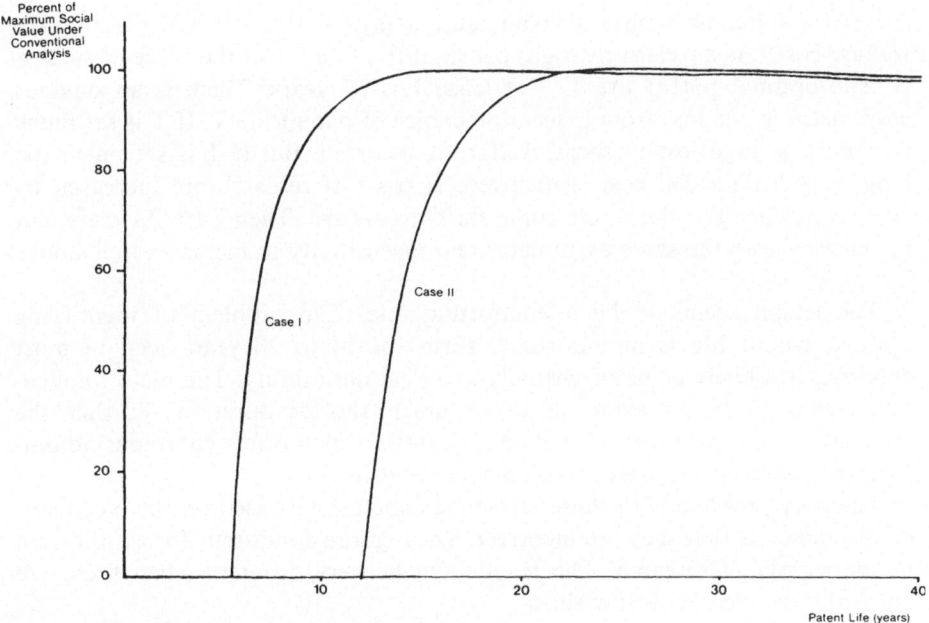

Figure 3. The Effect of Patent Life on the Social Value of Research: The Conventional Patent Economics (Parameters of Cases I and II are described in the text).

shorter period. In this analysis the optimal patent life, T^* induces a net private marginal return of $NMPB(T^*) \equiv (1 - \alpha(T^*)(MB - ML)$ illustrated in Figure 2. This equals marginal cost, MC, at a level of research m_2. At m_2, the social cost of reducing research effort by imposing a patent life, T^*, rather than awarding an infinite life is represented by triangle DEF. But the social gain from limiting the social loss, L, is represented by parallelogram MNDA. As T increases, the vertically hatched area MNDA increases in height and decreases in width. Optimal patent life is seen in this analysis as occurring at T^*, that T at which a marginal increase induces, via a decrease in m, an increase in the social loss triangle DEF owing to reduction in research from the infinite – T case equal in magnitude to the increase in area MNDA, the increase in the net social benefit of the remaining research, induced by shorter patent life.

In the type of analysis outlined here, optimal life T^* tends to be relatively long for typical parameter values. Furthermore, the maximum net social gain from limiting patent life at all tends to be small. This fact is reflected in the results shown in Figure 3, based on the model presented in Wright (1980), with constant marginal cost of research. In Case I, $r = 0.15$, $B = 1$, the value of a successful outcome $C(m) = 0.005m$, and $\mu(m) = 0.01m - 0.0001m^2$, and the welfare cost L is a (relatively high) constant fraction 0.1 of the value of success B. The optimal patent life T^* for Case I is 17 years. There is an obvious asymmetry in the loss from inaccurate choice of patent life T. If T is set much too short, a large loss in social welfare is incurred. But if T is set much too long, very little social cost is incurred. If costs of research are increased by 50%, as in Case II, , the whole curve shifts rightward. Then $T^* = 23$ years, but the curve shows the same asymmetry and insensitivity to increases in T above T^*.

The lesson seems to be a comforting one: The problem of identifying optimal patent life is unimportant. Terms of 14 to 20 years seen in most countries are likely to be of virtually identical desirability. The main implication seems to be to avoid terms of much shorter duration. Further the standardization of patent life within a country, which offers enormous administrative convenience, appears to come at negligible cost.

The only problem with these attractive conclusions based on the Nordhaus (1969) model is that they are incorrect. They ignore conditions for equilibrium in the supply of research. The results can be very different when these are correctly modeled, as I now show.

1.b. *The New Economics of Patent Life*

The conventional analysis recapitulated above treats the equilibrium supply of research under a patent system as analogous to the equilibrium supply of other competitive factors in a system which pays them their marginal revenue product. In Figure 2, if curve NMPB(T^*) represents for example the marginal revenue product of labor employed by a firm or industry, and the supply curve of labor is horizontal at a wage equal to OJ, then equilibrium labor use occurs under a wage payment system at m_2. Workers receive their opportunity cost, the wage OJ obtainable elsewhere, and they make no profit. Profits from production are represented by the triangle QFJ, and they accrue as rent to the owners of the "fixed" factors of production, land and/or capital.

In the market for research, however, if total research is m_2, the expected rent, $R(m_2)$, represented by triangle QFJ, is the expected net private value of the inventions or innovations produced, and it accrues to (some of) the suppliers of research. Now the results of research are by definition uncertain. Assuming for simplicity that all researchers are risk-neutral and equally proficient, the expected profit of any researcher i producing m_i units of research is $(m_i/m_2) \cdot R(m_2) > 0$. Researchers expect to receive more than their opportunity cost OJ per unit of research, so there will be an excess supply of research at m_2 units of research activity. The problem is precisely that of an open-access fishery (Gordon 1954) and other phenomena popularized as the "Tragedy of the Commons" (Hardin 1968). (For references to more recent literature see Dow 1984.)

The equilibrium in this situation is ilustrated in Figure 2. Equilibrium occurs at m_3 units of research, where (expected) private average revenue product of research, NAPB(T*) equals marginal cost, OJ. The average (expected) revenue product of research is the marginal private incentive. It is the expected return on the "research lottery" for providing a marginal unit of research. But Figure 2 shows that the marginal revenue product of research at m_3 is only OR, much less than marginal cost. In fact the total expected loss on the last $(m_3 - m_2)$ units of research is triangle FKL, exactly equal to the profits on the first m_2 units, triangle QFJ, so that researchers receive expected profits of zero, consistent with equilibrium supply.

In Figure 2 average social returns at m_3 exceed marginal cost by (S-J), even though net private returns are zero. If there were no patent life limitation, research would provide no social benefit in this example. The relevant net marginal social return, NMSB(T*) shows that too much research is supplied at much too low a net social return under the conventional patent life T* identified as optimal under the conventional Nordhaus-type analysis.

What is the remedy for this? One answer (for another, less practical, see Wright 1983) is to decrease patent life from T* to T**. As illustrated in Figure 4, which is a modification of Figure 2, a shorter patent life shifts the private average revenue product curve down to NAPB(T**) so that it intersects MC at point X, eliciting the optimal competitive research effort, m_4. Under this new analysis, a decrease in patent life has two effects on social welfare. It moves the net marginal social benefit curve NMSB up, increasing total social benefit for given m, and it moves m to the left, decreasing net social benefit if NMSB(T) > MC. The optimum patent life, T**, yields optimal research effort, m_4.

The good news carried by this newer analysis is that the maximum achievable net social benefit, represented by area ZXJ, exceeds the maximum net social benefit under the previously conventional analysis, area MNFJ, because the welfare loss due to the patent restriction is endured for fewer years. Patent life T*, optimal according to the previous analysis, is much longer than T**, the optimum under the new analysis.[1]

. 1. Note that at m_4, net marginal social returns NMSB(T**) greatly exceed net marginal private return NMPB(T**). The statement (Chapter IV) that "the extent to which social returns exceed private returns, reflects failure on the part of the protection system," is not supported by this analysis.

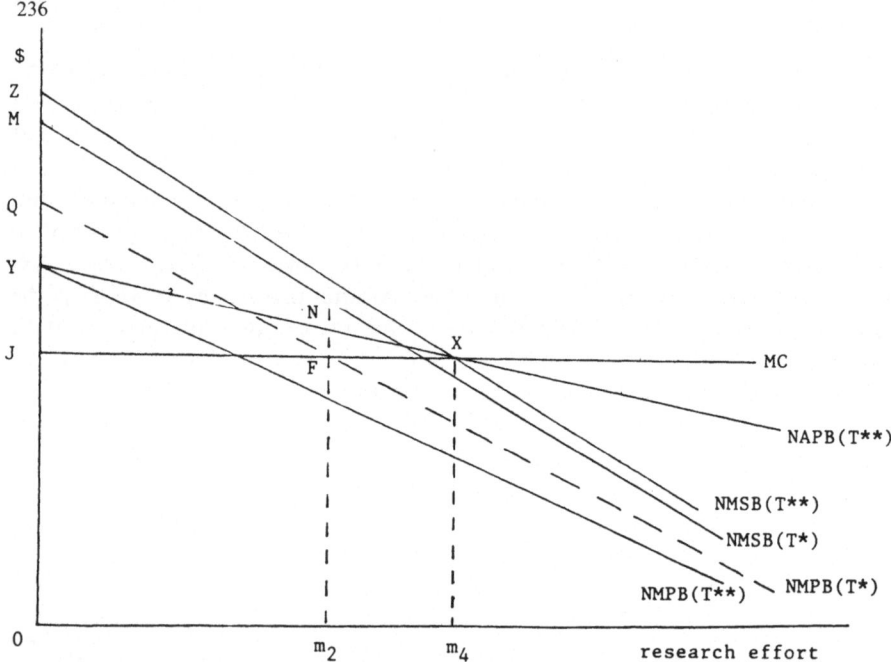

Figure 4. The Optimum Patent Life in the New Analysis.

The bad news comes in two installments. The first is that the optimal life appears in many reasonable cases to be much shorter than offered by existing conventional patent systems, if they offer full marginal appropriability. The second is that the optimal life is very sensitive to parameters that are difficult to identify, and small deviations in either direction can mean large losses of net social benefit.

All the above points are confirmed and illustrated in Figure 5, which shows the same two cases seen in Figure 3. The peak attainable net social benefit is in each case substantially above 100 percent of the peak illustrated in Figure 3, but mistaking case I for the higher-cost case II will result in an excessive patent life and the loss of more than half the net social benefits. The reverse mistake will result in a 100 percent loss, since such a short T will be chosen that no research is done at all.

This problem of sensitivity to the choice of patent life is less severe if the supply of research is less elastic with respect to incentives (see Wright 1985). But if supply is very inelastic the welfare loss caused by the patent restriction may not be worth incurring. Society may be better off without a patent system for such cases. Using an extension of this model to include asymmetries in the information held by government and the private sector about crucial parame-

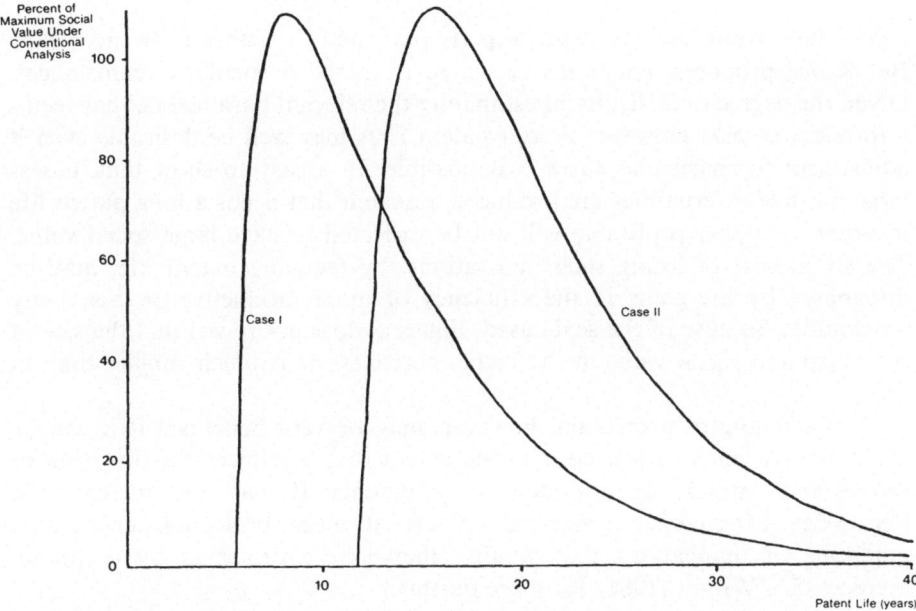

Figure 5. The Effect of Patent Life on the Social Value of Research: The New Patent Economics.

ters, I show elsewhere (Wright 1983) an example where prizes dominate patents if the research supply is inelastic. Direct contracting of research dominates both prizes and patents if the probability of success and/or the elasticity of supply is high.

If, as Kingston argues, the supply of innovation depends more on investment and less on the spark of genius than does invention, then the supply may well be very elastic, and the sensitivity identified here may be a quite important problem for patents. It is quite plausible that the probability of success is also higher for innovation than for invention, since the former is farther "downstream" in the process of research and development. A combination of high supply elasticity and high probability of success could indicate that direct contracting may be a favorable alternative to patents, but that issue is beyond the scope of these comments.

If patents are the chosen means of encouraging innovation, Kingston's argument for different lives for different kinds of innovation, with long lives associated with less profitable projects, is consistent with the analysis shown here. Under a patent system, the value of innovations is generally more fully appropriable than is the value of inventions. Where the inappropriable spinoff is negligible, there seems to be a good case for lowering the average patent life below the current levels (except for petty patents) of fourteen to twenty years. But Kronz proposes that terms be varied to match particular circumstances. Given the degree of difficulty in estimating the relevant parameters, that seems a formidable task. However shorter patent lives may well be desirable even if adjustment for particular cases is impossible. It is easy to show that, unless large positive externalities are produced, research that needs a long patent life in order to appear profitable will not be expected to have large social value. The social cost of losing such innovations, by reducing patent life, may be dominated by the gains in the efficiency of more productive or less costly innovation. Because of the scale used, Figure 5 does not reveal that the size of the maximum social value in the higher-cost Case II is much smaller than in Case I.

Strengthening of protection, however, may be very beneficial if it can be achieved efficiently at low cost, to the extent that it reduces the diversion of resources to attacks on and defense of patents. It may also increase the knowledge externalities generated, if the stronger protection encourages patenting of innovations that would otherwise be protected by corporate secrecy. (See Wright (1981) for more on this.)

In this section I have concentrated on the case of competitive innovators, that is, innovators without significant market power. These appear to be the focus of the Kronz and Kingston proposals. Lack of time and space does not permit analysis of patents under other market structures. There is an abundant and growing literature on this. Some references include Kamien and Schwartz (1975), Loury (1979), Lee and Wilde (1980), Dasgupta and Stiglitz (1980), Gilbert and Newbery (1982), Tandon (1984) and Reinganum (1981).

2. *Timing of the Award: Ex Ante* Versus *Ex Post*

Under the Kingston proposal, the Innovation Warrant is awarded before the innovation is achieved, on the basis of a review of the proposal which is subject to challenge only prior to the award. The Kronz Innovation Patent retains the current procedure of *ex post* patent awards. In this section I present some comments upon the implications of this distinction.

2.a. *Timing When All Potential Innovators Are Identical*

The fact that Innovation Warrants would be awarded before the resources are expended would mean that any waste arising from unproductive duplication of research would be avoided. It might seem, therefore, that the waste associated with competition for long-life patents discussed in Section 1 could disappear under the Innovation Warrant, without the reduction in patent lives discussed in Section 1. Because determination of appropriate reductions could be at best costly and difficult, Innovation Warrants might appear to have a significant advantage over Innovation Patents awarded *ex post*. In effect, the Warrant could be compared to a claim to a mineral prospect which confers subsequent exploration, development and exploitation rights. Indeed Kitch (1977) (see also Cheung (1977)) in his Prospect Theory claims that the current patent system for inventions protects the patentee's return on subsequent innovation by giving him exclusive control over the discovered technology.

Whether Kitch's conceptualization is consistent with current institutional reality is a matter of fact that has been disputed by Beck (1983). But in any case the theoretical implications are erroneous. If an Innovation Warrant or a patent prevented unproductive competition in subsequent research, it would still confer rents on successful applicants. If the supply of applicants is responsive to incentives, any expected marginal returns in excess of marginal costs must be dissipated in the application competition if a supply equilibrium is to exist. That is, the wasteful competition still takes place, but it is concentrated into the race to be the successful applicant. Kitch's "prospect theory," contrary to his interpretation, does not solve the problem of excessive competition. This is so even though the analogy with mineral prospects may be accurate, for the same problems should occur in mineral prospecting on public lands under the United States laws. (On private lands, private control of the land title prevents such waste, but a case could perhaps be made that the waste occurred in the competition for the original land title.) These phenomena can be viewed as special cases of the "rent-seeking" behavior analysed by Krueger (1974) and subsequent writers, including, prominently, Bhagwati and Brecher (1984) and Buchanan, Tollison and Tullock (1980).

As discussed in Section 1.b above, with sufficient information, the competitive waste of resources could be prevented by appropriate adjustment of the patent life.[2] But the point is that the *ex ante* Innovation Warrant is not inherently superior to the *ex post* Innovation Patent in controlling excess competition. On the other hand *ex post* awards may have certain advantages when researchers are heterogeneous, as discussed below.

(2) Alternatively, this waste could conceivably be eliminated by an auction of the contractual right to pursue a line of research as advocated by Yu (1981). A problem with the latter approach is that the original idea of pursuing the line of research may be the most crucial contribution. If the proceeds of the auction go to the state, the producer of such ideas receives insufficient incentive. If the proceeds are received by the source of inspiration, the excess competition problem reappears at that level.

2.b. *Timing of the Award with Heterogeneous Researchers*

A great attraction of the patent system is that it encourages researchers to make use of their own private information about their own abilities, the cost of research and the probability, timing and private value of success. Absent the common pool problem discussed in Section 1.b and the generally much less important welfare loss due to the patent restriction (illustrated in Figure 2), which cause the perceived marginal private value of success to diverge from the social marginal revenue product, these advantages over alternative incentive mechanisms would be overwhelming, as long as the government could not easily share all the knowledge held by private researchers.

If a patent system is to be used, and the common pool problem is tackled in a rough way by shortening the life of innovation patents that generate little inappropriable knowledge, it is obviously desirable to encourage private researchers to make use of their relevant private information in managing their resources. Conventional patents and the Kronz Innovation Patent (except for its misguided adjustment for capability, discussed below) do this by encouraging self-selection of researchers and independent choice of tasks.

But the *ex ante* award used in the Innovation Warrant scheme wastes private information about the ability of researchers to carry the task through to completion. For example, researchers A and B may independently hit upon the same idea. If A gets the idea one day before B, A receives the Innovation Warrant. Even if B is much more efficient in subsequent stages of bringing the task to market, he will be excluded by the Innovation Warrant possessed by A. (One might counter that A could hire B, but so could the government and, in either case well-known contracting problems might render this arrangement inefficient or impossible.) Society could be better off in an open competition in which B persists to the end and A retires gracefully upon (early) recognition of the differential in competence. The Kingston *ex ante* award encourages researchers who are quick with proposals but not necessarily efficient in the execution.

3. *Should Terms be Matched to Innovative Capacity?*

Both the Innovation Warrant and Innovation Patent proposals include provisions to vary the term of patent protection inversely with the innovative capacity of the applicant. In the Kingston proposal longer patent lives are offered for innovative investments that are larger relative to the size of the firm, and for innovative investments that are more risky. The terms of

Innovation Patents apparently would be adjusted to the innovative capacity of individual firms.

The motivation for adjustment of the term of Innovation Warrants for relative size appears to be to make all firms equal competitors in the innovation race, on the grounds that "[If] a patent is performing its work of protection properly, the result should be altogether independent of firm size." In the case of Innovation Patents, a similar notion appears to be extended to other dimensions of the characteristics of a firm; but this egalitarian stance does not appear to be justified. The correct criterion would be to aim for an allocation in which the appropriate social welfare function is maximized. Unless one wants to use innovation investments as very clumsy tools for income distribution, one would want to arrange research so that it is conducted most efficiently. This would mean that firms of the most efficient size will, through a process of competitive selection, obtain a disproportionate share of the patents.

Kingston claims that at present the very largest firms have a disproportionate share of innovations. I do not believe that this is true for the United States. If it were, and if it were due to higher "capability market power" (as he puts it), it would not be an inappropriate result, for capability market power appears to be synonymous with lower unit costs. If, on the other hand, innovation by large firms is favored by monopoly power stemming from less laudatory sources, there may be an argument for offering longer patent protection to smaller firms, as a means of reducing monopolistic concentration in the long run. If there is such an argument, protection should probably be related to the market share of the firm or some similar measure. Kingston's idea of offering government support against challenges could be an important encouragement for small firms, and could perhaps be limited to them.

However these proposals may not reduce market concentration, their clear pro-competitive intent nothwithstanding. Experience with the United States automobile industry, for example, shows that such industries can be made more competitive and more efficient by confrontation with international competition. If the proposals make it easier to exclude foreigners not engaged in local manufacture, they may be in effect instruments of protection from such competition that increase the welfare cost due to local monopoly as they reduce the gains realized from trade.

On the other hand, if discrimination in patenting procedures is avoided, it is by no means clear that the proposals would help small firms in the contest with larger domestic and international competitors. Under the proposed systems, large firms, with large international resources, might pre-empt any possibility of a small innovator exploiting his patent in other countries. The barriers to multi-country patent protection would now be much more severe for small firms, since they would include a requirement for the capital necessary to produce in each country. If the EEC is treated as one country for the purpose of this rule, then firms located there and in the United States will have increased advantage in innovation over firms from other developed and less developed countries.

In addition, it is not clear that different patent terms should be given to innovations in different risk classes, as proposed by Kingston. Much of this risk is eminently diversifiable via financial markets where the latter are well developed. In such an environment large firms do not appear to have any great advantage in handling risky projects. In fact in the United States the most risky research ventures may be unduly concentrated in small partnerships, due to the nature of the tax laws.

Finally, I have less confidence than is expressed in Chapters I-IV that modern accounting methods and panels of experts from the relevant industries can accurately and efficiently judge the prospects of innovations in order to set patent terms, and subsequently review past performance so as to make necessary adjustments in the workings of the scheme. If such expertise exists, it is certainly not common. Whether such a scarce resource should be devoted to panels of this type, rather than wholly applied to getting on with the task of private innovation, is a good question.[3]

4. Conclusion

The proposals of Kingston and Kronz contain elements which could substantially improve the patent system. Support by the state in defense of patents, especially those held by small firms and individuals, would help increase the ability of these groups, which do not have alternate sources of monopoly power, to obtain effective protection of their innovations from predation by others with superior financial resources. Especially if patent protection is made more legally secure, the suggested move to shorter patent lives would be broadly consistent with the insights gained from the new economics of patents which recognizes the common pool problem. In this new

(3) Analogous procedures have been proposed by Garnaut and Ross (1983) for evaluation by governments of appropriate resource extraction agreements for mineral deposits. Some success has been claimed in application of their proposals. But the implementation incurs large organizational costs, even when the parameters of the problem are well known relative to typical innovations. There is no possibility of applying resources of similar magnitude for every patent application.

analysis, the practical necessity of patent life uniformity incurs a greater cost than implied by traditional analysis, and the advantages of a patent system over alternative incentive mechanisms are less obvious. But if a patent system is used, the traditional conclusion that a uniform patent life should be long is essentially reversed.

Kingston makes a strong argument that the non-obviousness criterion for awarding patents is overly restrictive. Perhaps it should be subject to the test that if the elements of the innovation were available for a reasonable time, and if the patent is valuable enough to justify the cost of application, then the innovation should be considered non-obvious *per se*.

In the above brief outline of the analysis of optimal patent life, I have ignored much detail and many qualifications. In particular I assume that the entire marginal (but not total) benefit of the discovery is appropriated by the holder of the patent during the term of the patent. This conceptualization is at odds with that implicit in the prevailing legal view, which is that the consideration for the patent grant consists of the disclosure of the nature of the invention to others, as discussed by Kingston (Chapter 1). The patent is legally an exchange of information for protection of that part which can be embodied in the patented object. The implicit assumption is that not all the information can be protected by embodiment in the patented object, otherwise there could be no legal consideration in exchange for the patent right.

Curiously, any social benefit, in the form of consumer or producer surplus, generated when the marginal benefits of an invention or innovation are fully appropriated, is entirely ignored. The patent is not seen as a reward for research success, exploitation of which directly benefits (at least some members of) the public in comparison with the absence of such achievement. This legal stance could be interpreted as indicating that the legal tradition has understood that most of the social value accrues to the successful inventor, as in our "run of the mill" case in Figure 2, and that competition will tend to compete much of this surplus away under the current system, as outlined in my description of the common pool problem. If so the lawyers have been uncharacteristically modest in their discussions with economists on this topic.

I conjecture that the legal interpretation of disclosure as consideration may have encouraged the implementation of the "non-obviousness" criterion. Any invention or innovation that fails this test presumably generates very little new knowledge not embodied in the patentable object; therefore, according to the legal view, it offers the public no consideration in exchange for legal protection of the rights of the patent-holder. This point is generally more important for innovations than for inventions; more "applied" inventions are less likely to generate disembodied knowledge.

But the modern analysis presented here suggests that when patent life is optimized, a major part of the consideration is the net social surplus associated with exploitation of the embodied knowledge initially covered by the patent, in the years after it expires. The "non-obviousness" criterion is not needed to fulfill any requirement for consideration.

To the extent that the Kingston and Kronz proposals imply, on average, a shortening of patent lives for innovations, they receive support from the above analysis. However two caveats are in order. First, the Kronz proposal that inventions be denied patent protection entirely is completely contrary to the above analysis if, as is generally agreed, inventions offer, in general, more of the inappropriable knowledge externalities than do innovations. The correct economic analysis would imply that patent protection should be lengthened and/or strengthened, *ceteris paribus*, as the fraction of benefits that can be appropriated decreases. The second caveat is that the shortening of patent lives supported by the analysis does not take into account any strengthening or broadening of protection. Broadening patent claims increases the welfare loss at any life, and would imply further shortening of an otherwise optimal patent life.

As an invited critic, I have naturally devoted more than proportional space to those elements of the proposals with which I have difficulty. The *ex ante* nature of the Innovation Warrant award appears undesirable, and the efforts to adjust the term of the patent to specific cases, especially the more detailed deliberations implicit in the Innovation Patent proposal, appear very costly if not infeasible. Finally any attempts to make awards inversely proportional to efficiency-based innovative capacity are misguided.

It is easy to focus on these objections if one does not remember the great deficiencies of current patent systems. Kronz and Kingston have independently come up with proposals which may lead to signal improvements in the protection of innovations. Their efforts are to be highly commended.

NOTES

Barzel, Yoram, "Optimal Timing of Innovations," Review of Economics and Statistics (August 1968), 348–55.

Beck, Roger L., "Competition for Patent Monopolies," Research in Law and Economics Vol. 3, 1981, 91-110; "The Prospect Theory of the Patent System and Unproductive Competition," Research in Law and Economics Vol. 5, 1983.

Bhagwati, Jagdish, Richard A. Brecher, "DUP Activities and Economic Theory," in David C. Colander, ed., Neoclassical Political Economy: The Analysis of Rent-Seeking and DUP Activities, Cambridge, Mass.: Ballinger, 1984.

Buchanan, J., Tollison, R., and Tullock, G., eds., Towards a Theory of the Rent-Seeking Society, College Station, Texas: Texas A&M University Press, 1980.

Cheung, Steven N.S., "Property rights and Inventions: An Economic Inquiry" (mimeo, May 1977); "Property Rights in Trade Secrets," Economic Inquiry Vol. 20, 1982, 40–52.

Dasgupta, P. and J. Stiglitz, "Uncertainty, Industrial Structure and the Speed of R&D," The Bell Journal of Economics 11, No. 1 (Spring 1–80), 1-28.

Dow, Gregory K., "Pools and Streams: A Stochastic Model of Applied Research," Journal of Economic Dynamics and Control 8 (1984) 117–33.

Garnaut, Ross and Anthony Clunies Ross, Taxation of Mineral Rents, New York, N.Y.: Clarendon Press, 1983.

Gilbert, Richard J., and David M. G. Newbery, "Preemptive Patenting and the Persistence of Monopoly," American Economic Review LXXII (June 1982), 514–26.

Gordon, H. Scott, "The Economic Theory of A Common-Property Resource: The Fishery," Journal of Political Economy, Vol. 62, April 1954, 124–142.

Hardin, Garrett, "The Tragedy of the Commons," Science, 13, Dec. 1968, pp. 1243–1248.

Kamien, Morton I., and Schwartz, Nancy L., "Market Structure and Innovation: A Survey," Journal of Economic Literature 13, (March 1975),pp.1–37.

Kingston, William, "The Unexploited Potential of Patents." In European Economic Community Study Contract Pat. 1-83, "Direct Protection of Innovation" (Ch I above).

Kitch, Edmund W., "The Nature and Function of the Patent System," Journal of Law and Economics, XX, 2 (October 1977), 265–290.

Kitti, Carole, "Patent Policy and the Optimal Timing of Innovations," Ph.D. Dissertation, University of Chicago, 1973.

Krueger, A. O., "The Political Economy of the Rent-Seeking Society," American Economic Review, June 1974.

Lee, Tom, and Wilde, Louis L., "Market Structure and Innovation: A Reformulation," Quarterly Journal of Economics XCIV (March 1980), 429–36.

Loury, Glen C., "Market Structure and Innovation," Quarterly Journal of Economics XCIII, No. 3 (August 1979), 395–410.

Machlup, Fritz, An Economic Review of the Patent System, Study No. 15 of the Subcommittee on Patents, Trademarks, and Copyrights of the Committee on the Judiciary, U.S. Senate, 85th Congress, 2nd Session (Washington, D.C., 1958).

Nordhaus, William D., Invention, Growth and Welfare, Cambridge, Mass.: MIT Press, 1969.

Plant, Arnold, "The Economic Theory Concerning Patents and Inventions," Economica (New Series), Vol. 1 (1934), 30–51.

Reinganum, J.F., "Dynamic Games of Innovation," Journal of Economic Theory 25, No. 1 (August 1981), pp. 21–41.

Tandon, Pankaj, "Innovation, Market Structure and Welfare," American Economic Review 74, No. 3 (June 1984): 394–403.

Scherer, F. M., "Nordhaus" Theory of Optimal Patent Life: A Geometric Reinterpretation," American Economic Review (June 1972), 422–427.

Usher, Dan., "The Welfare Economics of Invention," Economica 31 (1964), 279–287.

Wright, Brian D., "The Economics of Invention Incentives: Patents, Prizes, and Research Contracts," American Economic Review 73, No. 4, 1983, pp. 691–707.

Wright, Brian D., "Is Protection of the Secrecy of Research Socially Desirable?" Paper presented at the Annual Meetings of the Eastern Economics Association, Philadelphia, Pa., April 10, 1981.

Wright, Brian D., "The Resource Allocation Problem in Research and Development," Yale Economic Growth Center Discussion Paper No. 353, June 1980, revised April (1984), in George Tolley et al, eds., Research and Development Policy, New York, N.Y.: Frederick A. Praeger, 1985.

Yu, Ben T., "Potential Competition and Contracting Innovation," Journal of Law and Economics, Vol. 24, October 1981, 215–38.

CHAPTER XI

André Bouju

The position of a practising Patent Attorney with experience of protecting innovation on both sides of the Atlantic Ocean constitutes a favoured observation post for appreciating the practical efficiency of the various patent systems.

From this viewpoint, it seems that there exists a contrast between the highly sophisticated rules governing the granting process of patents and the uncertainties of the technological and economic impact of the same patents.

Such a contrast reveals the inadequacies of the classical patent systems, which also exhibit signs of decay, of which the world-wide drop in Patents for inventions originating in the country of grant is only one: In France, for example, there were 17,000 in 1967, but only 11,000 in 1983 – a reduction of 35%.

Costs have also increased to a level which often puts them beyond the resources, not only of private inventors but also of young, small, innovative firms.

The time it takes to obtain a grant is in many instances incompatible with the rythm of innovation. This is now between 2 and 3 years in all the industrialized countries (at least 3 years for a Patent from the European Patent Office) which generally means that an innovation is not really protected during its launching stage, when it requires protection most.

Moreover, these figures do not take account of the delays which most systems permit those who do not want a grant to be made, to introduce into the proceedings by way of undue oppositions, which are regrettably frequent; not to speak of the time which litigation can take in case of infringement suits. In France, 5 or 6 years for a case involving technical expertise, is by no means exceptional.

Experience also makes one conscious of the harm which results from the double uncertainty attached to the actual legal value of a Patent grant. There is doubt not alone as to the validity of the grant in itself, but also as to the extent of its provisions. Such a situation surely cannot be satisfactory either to the recipient of the grant or to the public. Associated with this uncertainty in the present stage of most examination practices is a strong element of arbitrariness linked to patentability conditions such as "inventive activity".

Even the concept of novelty (which *a priori* would appear to be capable of being an objective standard) is treated legally in a way which seems unrealistic to those who are actually involved in industry. Why, these ask, should a Patent be rendered void by a document which is very unlikely to be discovered in the ordinary course of establishing the state of the art before engaging in innovation?

In my opinion a huge gulf has opened up between what business men expect of Patents, and the concentration on unrealistic minutiae which they actually encounter in the system. This reminder of scholastic philosophy in its last and worst days is poles apart from the factual precision of technical research and the means by which innovation comes about today. The scholastic analogy is reinforced by the way in which those who control the evolution of the Patent systems have excluded many new technologies (those concerned with information, for example) almost as if they feared invention and innovation in the case of their own institutional arrangements!

It is a law of nature that decline reflects inability to adapt to changing circumstances, and Patents are no exception. It must never be forgotten that their function is not only to distribute abstract titles defining in a more or less precise manner an intellectual "property", but essentially to "promote the progress of science and useful arts", as Article 1, Sect. 8 of the United States Constitution put it: in other words, to underwrite innovation.

The question of whether Patents are able to do this or not comes down in practice to whether or not they prevent imitation. The empirical evidence on this point is by no means reassuring, and is confirmed by my own experience in my practice (which I believe is no different in this respect from that of my colleagues all over the world).

To quote but a single study, Professor Edwin Mansfield, of the University of Pennsylvania, discovered that no less than 60% of innovations which were supposed to be protected by Patents, were successfully copied within four years of coming on the market. Even worse, the cost of getting his product on the market to the average imitator was only 65% of that of the average innovator. Similarly, copying took only 70% of the time which was needed for origination. "Inventing around" a Patent added as little as 7% to the cost of an imitator in mechanical and electrical fields, 30% in pharmaceuticals, with an average of only 11%. In the light of figures such as these, who could claim that the Patent system to-day is properly adapted to the purpose for which it ostensibly exists?

And yet, facing such technological "criminality", there is a fundamental need for protection of the kind which Patents are supposed to bestow. So much is this the case that we can perhaps say that if the Patent system did not exist, it would have had to be invented. Consequently, what is needed now, in fact, is invention of a juridico-economic kind, to *complement*, and not to replace, the conventional patent systems as we know them.

Chapters II and III of this book describe inventions of this type. What seems particularly noticeable about them, is that although one emphasises law and the other economics, they converge in a quite remarkable way. There is in fact a substantial area of identity between them.

The phenomenon of simultaneous invention is well recognised in technical fields, especially in the case of inventions "whose time has come," that is, when conditions are ripe for their successful marketing. If this applies also to social innovations, the circumstances may now be appropriate for introducing the provisions which are shared by both proposals and possibly some others, selected from either or both of them.

To reinforce this point about simultaneous invention, I propose to comment upon the proposals in terms of suggestions which I myself have developed. Some of these relate to the classical Patent system, but others could very likely be incorporated virtually without change in a system of direct protection of innovation of either type proposed above.

REWARDING EXPLOITATION BY "INCONTESTABILITY"

Any form of monopoly involves interference with the freedom of action of industry and commerce. In the case for classical Patents it is claimed that in exchange for monopoly, rapid and general disclosure of technical information is given through the published Patent Specification. However, there are good reasons for doubting the practical extent to which this actually takes place and its corresponding utility.

A more persuasive defence of monopolies of the Patent type is that they cover only *new* information. Consequently, there is no question of the public being deprived of anything to which it had earlier access for any desired purpose. As a justification of limited monopoly, therefore, the principle of novelty is objective (at least in theory) and acceptable to everybody.

In operation, however, difficulties inevitably arise. On the one hand, there are, in practice, substantial gaps in the information available at the time of grant; on the other, a grant which can be rendered void at any time through the discovery of some new item of information that is pertinent to the question of novelty at some earlier time, offers little mitigation of the risks involved in investment in innovation.

To reconcile the requirements on both sides, I submit the proposal for the classical Patent system, that *exploitation* should settle the question of novelty once and for all. That is, a Patent for an invention that is actually innovated would become *incontestable* as far as unobviousness and even novelty are concerned. No new information emerging later on could upset its monopoly.

This would give a *reward for exploitation*, in contrast to the sanction which used to apply in the past, of extinguishing the monopoly for "non-working". In other words, the general legal principle applied up to now of lessening the patent monopoly in case of lack of exploitation would be replaced by its reverse: a reinforcing of the monopoly strength in case of actual working of the invention. Reward is in many instances a better incentive than the fear of a sanction! In this respect the recent U.S. Patent Laws PL 97-127 and 97-414 implementing an extension of the patent duration if exploitation of the invention has been delayed by the need to obtain some administrative authorisation, is a good step in favour of reversing the conventional trend.

It seems that such an approach to incontestability of patents does not differ in any material way from that of both the proposals under consideration. All three link the strength of the grant firmly to exploitation, for which it appears to be a highly desirable, if not indeed necessary condition.

In a further refinement of this idea, I have proposed that the actual level of inventive activity demanded at the grant stage could also be materially lowered by exploitation. This was a further step in the direction of the originality criterion of both the new proposals, which eliminates abstract elements altogether, depending completely upon actual availability in the market.

Other ideas which I have been advocating, are directly complementary to the schemes for protecting innovation directly. One of these I call an *Innovation Protectorate* and it is perhaps best considered as a regime, or a general type of administration.

THE INNOVATION PROTECTORATE

General Background

For European countries, it is now beyond question that innovation is one of the outstanding means for curing the hardening of the arteries in a technological sense, from which they are suffering.

Two approaches to encourage the adoption of new techniques have been followed by national governments. The first is subsidy, which is interventionist and discretionary; the second, tax relief, is less discriminatory as between recipients. The effectiveness of both these approaches, however, is questionable, having regard both to the way in which they distort the market economy, and intensify economic inequalities.

It remains true that to a significant degree, innovatory activity calls for action to limit the risks faced by innovators, as far as possible. In the past, this led to certain ideas of protecting directly, ignoring the concept of novelty, which, as mentioned earlier, forms the ethical basis of classical Patents. The result of these ideas is protection for first exploitation of a new idea within a country, without requiring of the entrepreneur that he be the originator of that idea.

Such protection took the form of Patents of Importation, which under the name of Patentes de Introduccion, have been more common in Spain than anywhere else. There, anyone could obtain such a Patent, with a maximum term of 10 years instead of the usual 20, for the subject-matter of a Patent application by another in another country, even if this application had been published, as long as no exploitation in Spain had yet taken place. If the Patentee in Spain did not then exploit his Patent directly, or by transferring his rights, within three years, his protection would lapse. An important limitation to the monopoly granted by a Patent of Importation is that it covered only manufacture in Spain – whereas importation and sale remained free.

Although this system has been abolished by Spain since October 1, 1986, to bring its patent system into line with the European Patent Convention, it is clear that the importation patent principle copes with the problems associated with inventive activity and novelty in classical Patents by default, since what is protected cannot properly be called a new invention.

Like the proposals of Chapters II and III, Patents of Importation constitute a system of *innovation* protection. As such, they can be linked with stronger incentives to innovation and the investment associated with it, and such a combination forms what is to be understood by the term "Innovation Protectorate".

Institutional Arrangements

Two new agencies would be established to administer the proposed system:

The *Innovation Institute*, established by an individual State or by the E.E.C., which might advantageously take the form of an extension of the existing European Patent Office.

The *Innovation Stock Exchange*, a public institution associated with a country's existing arrangements for trading in stocks and shares. This could, of course, also be an *ad hoc* organ of the E.E.C.

The operation of the system would be as follows, with reference to the accompanying diagrams.

Any individual or firm (called the *Promoter*) with a technical idea which it is desired to exploit in a material way, would lodge an *Innovation Project* with the Institute, which would have to contain at least the first of the following five types of information:−

(a) A description and explanation of the idea whose exploitation is proposed, including information as to the state of the art.

(b) A financial assessment, with budget projections.

(c) A technical study, specifying requirements for production, sources of inputs etc.

(d) A commercial study, with comprehensive price data.

(e) Market reseach, including customer analysis and sales forecasts.

The project would then be studied and examined in conjunction with the Promoter by the Institute, which could call in specialised agencies of the State, to give their advice. If it is then considered meritorious, and if it had not already been exploited within the relevant territory at the date of application, it would be awarded "Protectorate" status.

This would protect the project initially for *three years*. If exploitation takes place within this period by the Promoter or by someone granted appropriate rights by him, Protectorate status is then extended until the *end of exploitation* or for *ten years* from grant, whichever is less. Failure to exploit within the first three years, results in the loss of protection.

As mentioned, the minimum content of an application which the Institute would require from a Promoter if the application is to receive consideration is (a) from the list above. If the Promoter is unable to provide items (b) to (e) himself, the project will be published, and interested Third Parties may then fill the gap on their own initiative, according to arrangements prescribed by the Institute.

If any gap remains unfilled, the project lapses. As a rule, 40% of the value of the Protectorate would accrue to the Promoter in recognition of his provision of item (a) and 15% to those who provide each of items (b) to (e). Obviously, if the Promoter provides the full list of reports to the Institute, he will enjoy the entire property and benefit from the Protectorate. Any Third Party who could prove that he had exploited the same idea before an application for Protectorate status, would thereby bring about complete or partial cancellation of that status. If his exploitation took place between the date of application and the date of grant of Protectorate status, he would be entitled to a free but non-transferable licence.

In addition to the usual rights under a Patent, the Protectorate also confers the right to approach the Innovation Stock Exchange for financing. Giving the Innovation Protectorate into Innovation Stock Exchange custody would automatically extend its duration to 6 years from the date of grant. The Exchange would be responsible for organising a public issue of shares in an Innovation Holding Company, formed to exploit the rights of Protectorate status. The value of the initial capital would be as specified in Study (b) but 25% of the shares would be reserved as payment for the transfer of Protectorate status to the holding Company. Although this Company would be effectively in being once its capital had been subscribed, its components could still be changed by negotiation within the Innovation Stock Exchange. Formation of the Holding Company under the aegis of the Exchange would automatically extend the Protectorate term to ten years.

The shareholders in the Holding Company would appoint a Managing Director by majority vote, and he in turn would appoint one or a number of industrial firms to undertake the actual innovation. These would either be minority shareholders in a Limited Company formed for this purpose, or licensees of the rights under the Protectorate, who would pay royalties to the Holding Company.

The latter could also use its resources to finance an industrial firm concerned with the innovation by loans of the conventional type, the interest on which would be added to the royalties. Protectorate status could not be challenged during the period of exploitation. In the event of failure of the project, the Holding Company would be liquidated and its remaining assets distributed in the usual way. Protectorate Status would come to an end and would thus have no residual value in a liquidation.

The foregoing is no more than a brief outline of this two-stage project, involving the Innovation Protectorate, and the Innovation Stock Exchange which depends upon it. The intention behind the arrangements proposed is, while remaining in the context of a free market economy, to encourage the development of technologies which have been slow to "take off".

Of course, the Innovation Protectorate comprises several rulings which may be considered to be arbitrary in nature and as such could be amended at will. On the other hand, it does not go so far into protection of commercial acts as the proposals of Chapters II and III which represent a much more elaborate corpus of doctrine.

But the merit – if any – of the Innovation Protectorate system, in addition to being compatible with the proposals advanced above, and of course with the conventional patent systems, is to reward those who gather all necessary information – and not only in a technical sense – to permit the practical and immediate exploitation of an industrial idea. This project, like the one which is now being studied, thus has as its sole aim that of being a direct incentive to innovation.

THE INNOVATION PROTECTORATE (1)

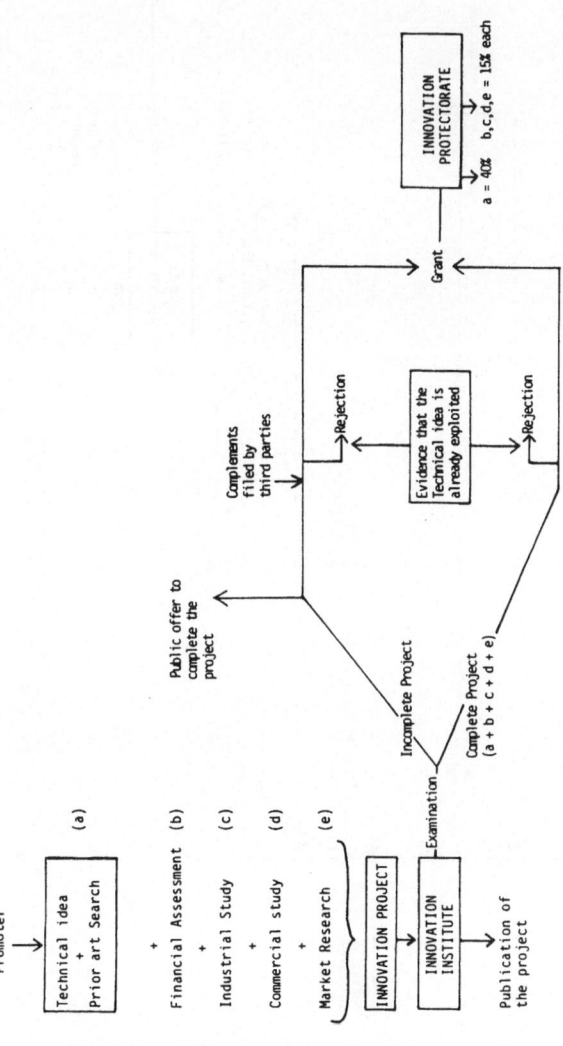

THE INNOVATION PROTECTORATE (2)

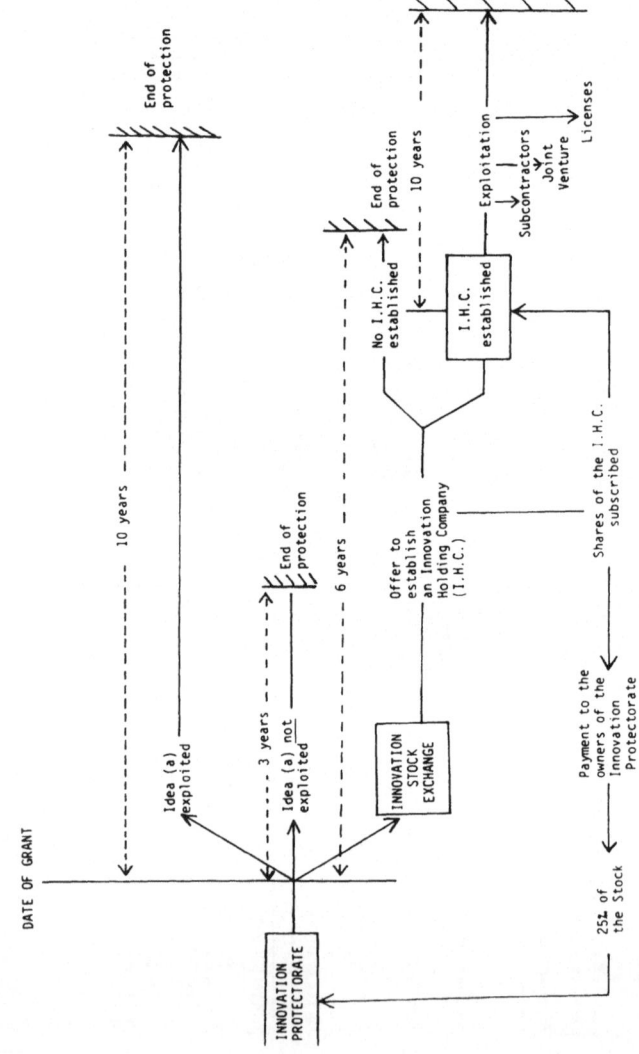

CHAPTER XII

Hermann Kronz

RESPONSE IN DEFENCE OF THE INNOVATION PATENT CONCEPT

For ease of understanding, I propose to respond to the comments of the various contributors in the order in which these appear above.

ANDRE PIATIER

Commencing therefore with Professor Piatier, I believe that his comprehensive definitions, diagrams and analysis of the innovation process are most valuable for refining and further developing the innovation patent. However, he appears to assume that, in accordance with the process by which an innovation emerges, the invention phase could be coped with by the patent of invention, the "pregnancy" phase by the innovation warrant, and reduction to commercial practice by the innovation patent.

The idea of having different types of protection in an innovation process is attractive, but is there a need for it?

The classical Patent of invention covers all three phases at present. It provides for protection of concepts (inventions) and innovations, that is, these concepts reduced to practice. The innovation warrant would do the same, through its emphasis upon a sort of advance protection for concepts in the initial phase. Likewise, the innovation patent provides for protection of concepts and innovations, i.e. concepts reduced to commercial practice, by means of the sort of *ex-post* protection it gives in the final phase.

The way in which the three systems of protection overlap, in spite of their (apparently) different ideological focuses, leads Piatier to suppose that they could at the same time form a sort of chain and exist in parallel. Unfortunately, this is not the case. The innovation patent system could only co-exist with the classical patent system, if the latter were to be legally subordinated to the former. I suspect that the same would be true in the case of the innovation warrant.

The reason for this is to be found in the incompatibility of the patenting conditions of the three systems. The kind of speculation undertaken by this author with respect to the three protection systems is – regrettably – undermined by the legal realities and parameters in each case. Piatier observes, at different points in his Chapter, that the object of the classical patent system was the protection of inventors.

I have no doubt that he knows very well that it is not the inventor who is protected by the invention patent, but this assumption leads him to think that the object of the innovation patent is to protect the innovator instead. This is not the case. The monopoly of an innovation patent will be granted to whoever (inventor or not) *innovates* first, instead of granting it to the one who *invents* first, as in the U.S.A., or *applies* first for a classical patent, as is the case in other countries.

The author pleads for dropping the inventive step criterion, and indeed, this is fundamental to both innovation patents and warrants. It seems to me, however, that all the suggestions for substitution which he cites are incompatible with the classical patent law's doctrine. These arguments are unable to underwrite a genuine replacement of the concept of inventive step, and therefore only succeed in re-naming it. Reasoning at this level of abstraction leads to conclusions which cannot really help to solve the problems.

MACDONALD AND MANDEVILLE

In contrast to Professor Piatier, who would like to see the classical patent system reformed, Drs. Macdonald and Mandeville consider it to be past redemption.

Their boldness in condemning it out of hand, recalls to mind Machlup's ironical advice, that if there were no patent system to-day, he could not recommend that one should be established; but since there was a patent system, neither would he recommend that it should be abandoned. (His comments referred to the U.S. patent system, but he was arguing in global economic terms).

And I also find echoes of my own writings about the pros and cons of the classical patent system in what these authors have to say. For example, in my early article of 1976 on the model for an innovation patent, I argued that "there is no causality between patents, specifically patent protection, and innovation..." and that "protection and innovation run parallel at best..."

According to Macdonald and Mandeville, the new proposals are for nothing more than "the existing patent system with nearly all its underlying and mistaken assumptions. Not only do their authors adopt the simplistic - almost static - view of the way innovation happens", they claim that we also neglect the fact that "innovation protection must impose some social cost..." And finally, they argue that we have done no more than "embark upon the relatively simple and pragmatic task of improving the effectiveness of intellectual property rights as incentives for innovation".

My only concern when conceiving and devising the innovation patent, has been to transform the classical system in such a way that the two statements quoted above (lack of causality and parallelism between patents and innovation) would no longer be true or at least would be less true than now.

The proposals therefore represent a pure theoretical exercise intended to show that if one wishes to maintain the patent protection approach, and if one wishes to make it more relevant for the promotion of innovation, then there is a solution.

The solution, however, goes far beyond the traditional scope of reform. This point, I think, has not been fully understood by Macdonald and Mandeville.

I must say, however, that there is nothing in their arguments about the role, production and use of information and knowledge, and of their specific qualities, to which I would not subscribe. I, too, believe that the classical patent approach of trying to pin down or fix the contents of information or "technical teachings" has no longer any place in the information era which is now developing.

I even doubt whether the protection of the form of information by copyright (referring to a "work") can be maintained in the long run. And that is precisely the reason why the proposals for the innovation patent move as far away as possible from trying to protect information in terms of its defined contents, which, of course, the classical patent system needs to do.

The fact that Macdonald and Mandeville did not seem to be aware that they are "pushing on an open door", suggests that this point needs further explanation.

The classical patent protects information, specifically technical teachings, as such. It tries to protect the contents of the information by making this into property.

In contrast, the innovation patent protects information as physically embodied (or incorporated or expressed) in the object of innovation. It is true that the classical patent protects information as embodied, but in addition, it also protects "disembodied" information. Reflecting the fact that it protects only information as embodied, the innovation patent forbids copying of the object of innovation in terms both of its contents and its form.

The innovation patent is no different from the classical patent when protecting an embodiment.

The difference arises because the innovation patent is for an object of innovation (in which absolute novelty and inventive step play no role), whereas the classical patent is for an object of invention (wherein absolute novelty and inventive step are decisive criteria).

The innovation patent protects things done (objects, that is, commercial goods and processes) and only in the specific way (contents and form) in which they are "done".

The classical patent protects things thought (ideas, "technical teachings", or rules, expressed in terms of claims) and in all imaginable ways of corporeal expression (as long as these expressions remain within the scope of the "teaching", rule or claim).

Whereas in the classical patent system the justification for granting a monopoly is that there is property in a technical teaching disclosed to the public, in the innovation patent, there is property in the specific execution of a technical teaching so as actually to bring into being a commercial object (or process).

The premium (patent) is for the innovative (that is, risk-taking) act; the premium of the classical approach is for the finding of a new technical rule (information).

In the innovation patent "new" does not refer to the teaching, but to the "act". This "act" (object, product or process) must be new within the country, whether the underlying technical teaching (information) was new or not, inside or outside the country. The monopoly is granted to the first innovator (doer) not to the first inventor or applicant (thinker) in respect of a technical teaching.

Consequently, an innovation object, when patented, is, legally speaking, only little different from any other commerical object, because both incorporate specific information, of which the value, for the innovation patent, consists only in the fact that it is reduced to practice in the respective country for the first time. It is therefore wrong to suppose that innovation patents could in any way sequestrate information or interrupt or hamper its free flow or use.

From all this it should be clear that the understanding of "strong property rights for technological information" which underlies the innovation patent, is quite different from the understanding of the classical system. In addition, there is no risk that "those (Patent) rights might well actually stifle the overall innovative process...", as Macdonald and Mandeville fear.

In the innovation patent, "strong" refers to the fact that once granted, it cannot be revoked or declared void, except in the case of fraud in obtaining it. An innovation patent is "strong" in that arguments like "the concept of the object is within the aptitude of people skilled in the art" are irrelevant. But it is "weaker" than a classical patent in that it has only a reduced scope of protection. It protects against copying, including copying by the use of "technical equivalents".

It also protects "optional" concepts disclosed by the patentee before the grant of the patent. "Option" claims, however, do not at all impede the working of their subject-matter by third parties, because they carry an obligation to grant a licence.

The main purpose of the concept of option claims is to cope with the problem of infringement in cases where the supposedly infringing embodiment of the object of innovation is no longer only a "technical" equivalent, but a "conceptual" one (a "legal" equivalent).

Whereas, in the classical patent law, this problem of equivalency is the subject matter of litigation procedures (the outcome of which is almost always dubious) in the innovation patent the question of infringement can only arise if the subject matter of the option claims is "worked" without licence. And since granting a licence is obligatory, there should be no, or very little litigation.

The whole construction of these provisions is guided by the principle that as much innovation as possible should take place. It is of more value to society that an innovation should be carried through, than that it should be prevented by the restrictions of monopoly claims.

It is difficult to understand, therefore, why Mandeville and Macdonald think that the innovation patent "would neither bring forth more innovative effort from successful high technology firms" nor could it "for long, instil new vigour in firms that have failed".

The first-to-innovate principle should exert upon everyone who wishes to take part in the race for a patent, a constant pressure to act and to contribute towards realization of the innovation object.

It is worth while to race for an innovation patent since, once granted, it cannot be invalidated. Further, it is not a premium for scientific and technical intelligence, but for having brought a product, device, process, object, etc. into first use in the national market. Those who do something new for customers are to be rewarded by a (limited) exclusive right allowing for the recovery of their investment plus a decent profit.

Thus, in Senegal, this "something new" may be a simple pump that is already available only in Brazil, for example. The first to introduce this pump into Senegal will obtain an innovation patent, if he wishes. As for Silicon

Valley, no firm should be able to get an innovation patent for a high-tech concept until it has been reduced to practice.

What about the charge made by Mandeville and Macdonald that the innovation patent is based upon the (mistaken) linear model of the innovation process?

This fails in the first instance, because the innovation patent has nothing to do with "invention". Among the objects which it protects, there may indeed be inventions in the classical sense. This does not matter. The only question is: Is the object brought into public use for the first time in this country by the applicant? The innovation patent is concerned only with objects (including processes) of a commercial character.

In the classical patent law, "technical teachings" (being the solutions to particular problems) are protected, but the reduction to practice of many of these teachings does not make commercial products or processes of them. For example, an electronic device for producing a particular shape of control signal may be patentable under the classical law, but may never be fit to be sold as a commercial article.

Innovation patents would protect entire articles, whether they contain many different "inventions" or other heterogeneous concepts. The "entity" that is to be protected is the article (product or process) as offered for sale or other commercial use. The principle of "unity of invention" in the classical Patent system is replaced by "unity of the goods".

Another reason why the innovation patent has nothing to do with the "invention phase" is that it provides for publication of findings, "inventions", teachings etc. even before filing a patent application, and without giving any power to these disclosures to destroy novelty; yet, for those same disclosures, an innovation patent can and will be granted to the first one who reduces them to practice in the country, for commercial purposes.

As to the "diffusion phase", "public use" is not conditioned by a minimum of diffusion of the object into the market. However, the innovation object when lawfully brought into commercial use, must at some stage lose its patent protection (in a way analogous to "consumption" of the patent in the classical law).

The innovation patent's preference for protecting "the tangible bit of embodied technological information" is motivated by:

(a) the requirement of conceptual rigour in its formulation; and

(b) the problem of proving infringement.

As to (a) a process or method, when formulated in a patent claim, is little more than a technical teaching of things and acts that shall be done, and is thus the classical way of claiming "ideas".

In a process, the innovative act only takes place when the process "works" (the process-in-being!). This leads therefore to the notion that a process should be expressed, as a process-by-product (or by objects). In order to show how this could be done, the example cited by Macdonald and Mandeville was given.

Of course, tradition and inertia may be expected to keep the process-approach of classical Patents alive. As to (b) it is clear that, in a patent, for instance a specific heat treatment of material, the terms of the precise apparatus's media, settings and sequences must be expressed in the claim.

Proof of infringement of the patent may then not only be derived from the product, but also from the knowledge of the existence of these media, settings and sequences in the apparatus of a "suspicious" enterprise. This may help considerably in "discovery actions".

Macdonald and Mandeville also charge that direct protection of innovation would mean that the "patent system is extended unnecessarily in the development of the information economy". What is proposed is indeed "converting information from an inclusive public good to an exclusive private good".

However, this information is "innovative". Only that part of the total, for which there is demand (i.e., which justified investment and risk) only in the specific country (the domestic market) and only one very specific corporeal form of expression, is given protection.

It appears to me that the innovation patent procedure of "step by step" disclosure has not been fully understood.

Firstly, no type of "non-embodied" information dissemination is detrimental to later patenting. It does not destroy novelty and it does not "anticipate". This concept is fundamental to the innovation patent and totally contrary to the approach hitherto used in classical Patents.

It is stressed that the monopoly goes to whoever innovates first, not to the one who thinks, publishes, invents or applies first. Yet, a useful legal effect is attached to the date of filing of information in the Patent Office concerning an entirely new conception. This effect regarding the scope of the right is explained in the diagrams in Chapter II above, and additional details with regard to the effects of protection of the filing are also explained.

The innovation patent is only granted when the object is ready for, and put to commercial use, that is, just at the moment when all the information necessary for full disclosure in writing is available. There is consequently no requirement for a "full description of the invention" to be deposited at the start of the patent application procedure, as there is the case of the classical patent law.

Instead, there is provision for depositing information (items of description which can form part of the later "patent specification" and of the claims) step by step at the Patent Office, and certain legal effects are linked into these depositions.

Of course, such an approach carries certain dangers with it, since competitors may take advantage of it, and be quicker in reducing the disclosed information to practice. But that is just the purpose of the innovation patent – to promote innovation. The innovation patent system is not a crutch for entrepreneurs nor a machine for distributing gifts to intelligent thinkers. It exists for the purpose of promoting innovation and competition.

I turn now to the section entitled "practical problems".

Macdonald and Mandeville doubt the value of the concepts of the innovation office and of the variable term of protection. In the light of the explanations in the preceding paragraphs, it should be clear that the tasks and way of operation of an innovation office would be very different from those of the classical Patent Office.

"Examination" would not be concerned with scientific and technical issues but with economic ones. The objective of the Office is the promotion of innovation, not the generation of intellectual property. Patent information and documentation would assume quite new and different purposes and connotations, from those they have at present.

Would it lead to a big bureaucracy?

I do not think so, since firstly, much of the information required can be obtained from existing sources and services.

Secondly, although the number of applications for innovation patents could well be higher than for patents for invention, the number of actual innovation patents should be much lower. Since "examination" only takes place when proof of introduction into commercial use of the innovation objects can be produced, and since this in turn depends on many economic factors of a risky character, the workload of the Innovation Office should be smaller than that of a classical Patent Office.

But even if the load were the same, the actual work would be of much more social and economic value than to-day. Probably not more than 10% of classical patent cases would be granted a patent of innovation. Perhaps a further 10% of the cases refused under the classical patent system would be granted a patent of the new type.

In this context, it should not be forgotten that the more innovation objects are imported into a country, the fewer patents for innovation will it be possible to grant, unless patents of importation are asked for. The terms of the latter would be short because of the limited scale of the "investments" needed to bring an imported object to market maturity, as the new system requires.

At any rate, for the first time in economic history, there would be no patent in existence which is not "worked'! This must have a significant impact upon competition policies based on industrial property, as well as on other policies. It allows intellectual property doctrine to be reduced for the first time to its minimum area of utility. Consequently, there is no need to expect a "monstrous bureaucracy" to develop.

Determining the duration of protection is easy, if figures by category (e.g. of sales per product class) are readily available. Some such statistics have to be collected in any event as a result of law or regulation. The task is less easy if scientific market research is necessary. A proper book-keeping system for innovation investment could help a lot, as could tables about the frequencies of innovative change in the various branches and categories of product and processes.

I do not believe, therefore, that in view of the very specific (narrow) scope of protection offered by an innovation patent, taken together with the increasing frequencies of technological change, there will be much pressure for long monopoly terms. Long terms could be obtained where needed, by co-operating in a joint venture or project with a partner to whom the new system gives the right to a long term (e.g. the private inventor or small entrepreneur).

The variability of the patent term can best be considered as a technical problem within the area of competition policy. And the value of a patent for innovation as an instrument in connection with regional or sectoral development policy, structural policy, R&D policy, etc., should be stressed. The reason will be clear when it is understood that the classical Patent is inescapably linked to the doctrine of property.

There is a strong case that property should not be manipulated according to the circumstances of people, nor should it be used to discriminate against them. The patent of innovation responds instead to social and national interests in terms of economic development. It may therefore be tailored to these interests as they evolve.

To give an example: One could not argue against a decision by Mali not to provide for patents of innovation for pharmaceuticals but to do so for irrigation installations, nor against an Australian refusal to grant patents for mining equipment. Such restrictions might easily be seen to be in the national interest. But if both these countries decided that they would no longer grant classical patents in these areas, trouble could be expected, since this would come close to denial of property rights.

In order to allow co-existence of classical patents and of patents for innovation, there would be need for a provision whereby patents of innovation cannot be invalidated by virtue of patents of invention covering the same subject-matter, as long as these latter patents are not worked. But in the long run one of the two systems must be dropped.

GORDON TULLOCK

Turning to Professor Tullock's Chapter, I welcome his support for the idea of extending the patent system to "as many kinds of intellectual property as possible". However he then refers to a listing in relation to the innovation patent entitled "Filling gaps left by classical patents", which he thinks specifies "a number of areas where... clearly we should have such protection".

For the sake of intellectual honesty I would like to clarify this point, so as not to take credit from an interpretation by the author that may be based on a misunderstanding of this list. In fact, this refers to a number of types of inventions (not areas) where very often the lack of inventive step (non-obviousness) leads to the rejection of the invention by the patent offices.

Now, since in the innovation patent the criterion of inventive step is no longer applied, it is clear that many more "inventions" become "patentable". But this only happens when they are reduced to commercial practice, which cuts down the number of patentable cases drastically.

There is, however, another aspect of the innovation patent that actually corresponds much better to Tullock's interest in expanding the system of patent protection to more areas. This is the fact that it does not protect "concepts" (called "inventions") but only physical embodiments of concepts (products and processes) of a commercial character.

In the classical Patent system, objects like living organisms *per se*, computer programs, games, plans and schemes (e.g. in management, book-keeping, teaching) etc. are not patentable because classical patent theory refers to the old concept of a purely "physical" world.

In contrast, these can be protected under the innovation patent system, because this does not distinguish one type of "invention" from another. It only asks, can these objects be sold or used in the commercial world, and are they "innovative"?

Professsor Tullock refers to the problem of patents having a scope of protection which is often either too small or too large. He also mentions the role which the concept of copy-claims and option claims can play in this context to help in solving the problem. Though conceived for a patent system for the protection of innovation (of commercial goods and processes) this concept could be used also in the present patent system for the protection of invention, under the following conditions:

(a) only one patent claim, restricted precisely to the preferred embodiment of the invention, would be allowed (this is the copy claim).

(b) a number of additional claims, referring to variants of the preferred embodiment of the invention, could be added (these are the option claims).

If these conditions are respected, there would always be the possibility of developing the scope of claims from "restricted" to "large" and vice-versa. Thus, the patent office would never accept a "large" copy claim, but may allow extension of protection by means of "option" claims; or it may not allow (or reduce in number and scope) these option claims. Of course, even this slightly modified application of the option/copy-claim approach, to the classical patent system, constitutes a major ideological change and may be expected to meet with resistance.

The comments of the author on the proposed new type of patent office, confirm my impression that his conclusion in respect of the proposal for a new patent system is that we should not make a radical change. Rather it is that we should pick out of the proposals some useful concepts for adapting the existing patent system to present and future needs.

Professor Tullock gives this impression by his insistence that the problem of proper protection and promotion of innovation is not "the most important problem we face in present-day patents". He thinks that the problems we have can be solved if the present patent law is properly enforced.

Undoubtedly, there is some truth in this, but all of the "reforms" of the patent system that have been undertaken up to now, have only worsened these problems. That is why I was forced to the conclusion that we need a fundamental reform, questioning even the very concept of intellectual property!

Let me take the author's example which concerns obtaining "know-how" from the patent. I think the reason why "know-how" is lacking in the patent specification, is that the doctrine of "inventive step" does not permit the protection of a craftsman's practical knowledge and rules. So why should an applicant disclose it in his specification, since he will obtain no protection for it in exchange?

But if the "inventive step" criterion is dropped, as a necessary move towards protecting "know-how", it will then be difficult, if not impossible. to sustain the concept of "invention", and even that of intellectual property itself. Indeed, their abandonment is substantially what the innovation patent system proposes.

Alternatively, take the problem of determining what is infringement of a patent. As long as we allow "large" claims and consequently the inventive step criterion, we will have this problem. If we allow the "copy" claims and limit infringement strictly to the specific contents of "option" claims, we reduce the scale of the problem drastically.

I do not accept that later innovators who develop embodiments of the invention, of which the original inventor or patent holder has never dreamed, should be blocked through the instrument of large claims. For this reason, I am unable to agree with this author that "it seems to me the switch from actual use or working of the invention to permitting people to simply file ideas, has been an improvement". Indeed, I think that the source of all of the problems we have in industrial property to-day, especially in patent protection, is to be found precisely in that switch.

Professor Tullock appears to believe that the innovation patent requires "the person receiving the patent to actually go into business before implementing his idea". This is a misunderstanding. What is demanded is that, before the patent is granted, the object of innovation is in the stage ready for commercialization.

Nor is it strictly true that the innovation patent cannot be given to someone who has only a "new idea'. He can get the patent the moment he proves that he (if necessary in co-operation with another) has reduced the idea to commercial practice.

It is perhaps necessary to recall here again the fact that the innovation patent is granted, not to the first inventor, nor to the first applicant for a patent, but only to the first innovator. Therefore, compared with any classical system, it constitutes a radical change in basic philosophy.

It can fairly be claimed that it is able to "stand on its own merits as an improvement in efficiency", as Professor Tullock confirms. And its efficiency as a means of dealing with Japanese strategy and ideology is striking. Needless to say, I greatly welcome this author's suggestion that the proposals for such radical change should be tested in limited areas.

AUBREY SILBERSTON

I am grateful to Professor Silberston for his very clear summary of the essential features of my model, which precedes his assessment of these. At least as far as the technical features are concerned, however, I doubt if the problems he has found in relation to them, really exist to the extent he believes. The main problem lies, in my view, in the well-known unwillingness of vested interests to accept change.

It is of course wise counsel to say that the direction in which (we) should move is that of "attempting to modify the patent (and copyright) system." I think, however, that the margin for modification, at any rate of the patent system, within its present boundaries, is now exhausted. The classical patent will therefore continue to lose ground, and will be able to deliver progressively less of its proclaimed potential. Professor Silberston suggests that some minor modifications of the classical patent system are all that are "most likely to occur in the foreseeable future." However, if anything falls into the category of predictable future events, it must surely be the continuation of the long-established trend for the classical patent to lose economic significance.

It is to reverse this process that I ventured to devise my model of protection. So far, of course, this is only a blue-print, not even a prototype. What I have in mind is therefore more than "a number of aims." It is a real change, or at least an attempt to get the fundamental concepts of the classical patent put on the "test-bench." There, they would be checked against the proposed models in order to find out if what we have at present is really the best system. We badly need to know whether all that it needs is to be "modified", so as to be brought into line with the constraints (in terms of "pace" and not only of "substance", because the problem is a dynamic, not a static one) of the "normative power of the facts." Carrying out this exercise on a larger and deeper scale than I have been able to do, would, I think, be a service to the institution of patent protection.

H. W. DE JONG

A different type of objection to direct protection of innovation is to be found in the Chapter by Professor de Jong. He argues:–

"It seems to me that to establish whether a product or a process is new is much more difficult than to verify whether an idea is new".

Under the classical patent law, the novelty of an idea is examined mainly by a literature search. This means that searching the novelty of a product or process is virtually identical with searching the novelty of an idea. In the innovation patent system "novelty" is based exclusively on "domestic public prior use" of a product or process available commercially.

It is true that proving "prior public use" is more difficult than proving prior publication of the idea in the literature. But this is just the point. I do not think that a society interested in improving its lifestyle through innovation, is interested, in the first instance, in ideas *per se*.

Progress is brought about by acts subsequent to ideas, by doing, by …innovation, which is, for my purposes, simply the production of novel products (products with no domestic prior public use) and the implementation of novel (in the same sense) processes and methods.

If domestic prior public use is the only criterion for patentability, then there will be much more awareness in the business world, of new products and processes. In the long run, this should give us the documentation necessary for establishing domestic prior public use through searching the "literature" (instead of through enquiries); and amidst this literature, there will be the patent specifications of many patents of innovation.

For the first time, patent specifications will always refer to "worked" concepts of ideas or inventions. So, the fact that at present novelty of ideas may be more easily established than novelty of innovations, is no argument against the innovation patent.

Professor de Jong also wonders how far innovation patents would be preferred to invention-patents, and thinks that the numbers will only be "somewhat reduced". I believe that the reduction in numbers could in fact be quite large, on the following grounds:–

Before an innovation patent system can be implemented, we need an inventory of products and processes (innovation objects) that are used, or have been publicly used in the country. There are of course no comprehensive catalogues of such objects available now, but preparing them would be a valuable exercise in itself.

A product or process would then be "novel" only if it did not figure in this inventory or if at least its novel characteristics were not among the qualifications described in the inventory. Essentially, we would be concerned with "modifications" of something which already exists, so that innovation patents would be, in number and "level", much the same as utility models. However, since the term of protection is essentially determined by the investment to be recovered through sales, etc., all cases involving only small investment will not be worth protecting because the term will fall within the frequency of innovative change.

In addition, objects requiring very big investment will only rarely be protected. This is because the innovation patent requires that, if the novelty concerns the whole construction of those objects, proof must be given that the entire object has been brought into the commercial stage at the time that the patent is granted. If the novelty refers to a detail of such objects, the detail must be a commercial good in itself and ready for commercialization.

Further, novelty can easily be destroyed by importing the objects for which it is believed there is a demand in the domestic market. Of course, in this case the exporter or importer may look for a patent of importation. But is it often that the investment made by an importer would justify looking for its recovery through an innovation patent?

An exporter might not be allowed to use his investment twice, that is, to obtain an innovation patent in his own country (if legislation provides for it) and in the country to which he exports the goods. He might have to divide his total investment into "domestic" and "foreign" shares.

Since in the innovation patent, the factors which control competition, and not, as in the classical patent system, the novelty of ideas, are the dominant influence on the taking out of patents, I assume that these factors will lead to a substantial reduction in the number of patents applied for.

I think that this will be healthy for the economy and will reduce the burden of examination for patentability. For the first time, this will concentrate on economic issues such as innovative capacity, investment, risk and term of the patent, instead of items such as absolute novelty, inventive step, unity of invention and similar abstract exercises.

I agree with Professor de Jong that a "simple offer to sell... does not make a product marketable". But no such offer will be valid (for the Innovation Office) if there is not at least one innovation object ready for sale. And to this end, investment is needed beforehand.

Are there likely to be many entrepreneurs who would invest simply to produce a dummy innovation object in order to get an innovation patent (which would be invalid in any event because of fraud)?

I am not worried by the danger of granting a monopoly for an object for which "the market would not develop"; nor would I fear the grant of an "unduly long monopoly". This is because, firstly, the innovation patent has a very restricted scope of protection, and secondly, if the lack of demand kills the product protected by an innovation patent *de facto*, then, so be it. This is what also happens with classical patents.

The difference, however, is that the classical patent, by virtue of the large range of the protection it gives (encompassing very many imaginable and also unimagined embodiments) continues to "block others from introducing similar products" – which Professor de Jong fears the innovation patent might do.

But in contrast, *every* modification can obtain an innovation patent if it is neither a "technical equivalent" nor an object of an option claim of the initial patent – and in the latter case, a licence must be granted by the initial patentee.

This point is important because only with this provision can prior publication and diffusion of concepts be allowed, without losing the chance of getting a valid patent later on. The only condition is that of being the first actually to reduce the concept to commercial practice in the country (the "first-to-innovate" principle).

Professor de Jong's objections to the concept of the "substitute innovator" and to the "conclusion of a contract... about something which is as yet unprotected" do not take account of all aspects of innovation patents.

Firstly, certain legal effects are attached to the filing of information and claims concerning the innovation object, at the patent office; secondly, the absence of legal protection after or before filing is irrelevant, because if the concept of the innovation is reduced to commercial practice it will necessarily take on a specific individualistic shape according to the ideas, views and imagination of the innovator. His protection will cover exactly this and no more.

But within the relevant specific limitations, there will certainly be patent protection, notwithstanding early disclosure of the concept. Of course, the latter carries certain dangers for the applicant, as mentioned earlier, but these add to the incentives towards innovation.

This risk could of course lead to the practice of more secrecy and less disclosure. This does not matter, because if, one day, the innovation object goes into the market (hopefully protected) the purpose of the proposed patent law is still attained; if it does not, where is the disadvantage of the innovation patent as compared with the classical patent? What value can the ideas published and protected by the latter have, if they are not implemented? We no longer need a patent system for stimulating ideas, but R and D.

Although de Jong regards it as important, I am not sure how far we really need a definition of innovation. I myself use such a definition only for the purpose of making the innovation patent better understood. What we do need are definitions of what is a product, what is a process, what is technical equivalence (we already know a good deal about this) and how can we best give appropriate weights to acts of prior use.

These should all be approached with the primary objective in mind, which is to reward the risks taken by entrepreneurs in serving developing market demands or social needs, in a competitive environment.

I think that the Innovation Office will be able to distinguish "new" tangible objects about to be put on the market, and which are candidates for a patent of innovation, from given tangible objects already on the market. A definition of innovation is not an essential tool for this work.

"Negotiations between firms before the granting of a monopoly" are therefore not as "paradoxical" as de Jong thinks, because, as explained earlier, the firm capable of "innovating" a concept will certainly obtain an innovation patent for it, since the probability that the chosen physical embodiment will be identical with an object already on the market is effectively zero. So negotiation to involve a "substitute innovator", with the object of attaining the most profitable patent term, is perfectly possible.

I can readily understand that the way the private inventor is treated in the innovation patent system may be disappointing to many people. However, we have to face reality. No system of protection can provide private inventors with the money and qualifications necessary to manage an enterprise.

But what it can do is provide them with some advantage which improves their negotiating position when they are trying to get their ideas into commercial practice. And this advantage, in the case of the innovation patent, is the longer term of protection they obtain compared to firms.

The latter possess a higher capacity for innovating, that is, for reducing an idea or concept to commercial practice and introducing its embodiment to the market. Consequently, the Patent Office would have to assign a shorter term for the same innovation to a firm than to a private inventor.

Such a firm will therefore be much more easily persuaded than under the present patent system, to buy the ideas of a private inventor and involve him in their reduction to practice. This is because co-operation between them will certainly obtain a term for a patent of innovation that is the arithmetic mean of what either would receive on its own.

But, apart from reinforcing the negotiating position of the weak and small inventors, innovators and entrepreneurs, what I aim at is to put pressure upon *all* patent owners. The rules of the innovation patent make it clear to them that, if they do not respect and use the term of their monopoly to good effect, they will not be able to recover their investment plus profit before having to meet competition from imitators.

In the classical patent system, you may implement the patent or you may not; you can enjoy the maximum term if you like, and meanwhile you are blocking others from implementing ideas you will not use. I see no rationality in this, except respect for the absolute rights of property. But is this the way to look for society's endorsement of a Patent system?

I agree with de Jong that it may be "highly questionable" to grant patents for non-technological objects. I only wish to point out that the innovation patent is *capable* of conferring this type of protection, if it is considered desirable that it should be granted. The present patent doctrine renders classical patents incapable of protecting objects of this type.

I can readily understand the nightmare experts anticipate from having to deal with questions such as determination of the term of the patent, the "innovative capacity" of firms, the evolution of market demand, and so on. All I ask is acceptance of the point that if the innovation patent system were as long in force as the classical patent system has been, we would be able to answer these questions reasonably.

Even as things stand, we are not without information to which reference can usefully be made. For example, we know for many given products, *ex-post*, how much was invested, how much was sold, how long innovation took, etc. which should enable us to devise fairly good measures for the monopoly term.

Nor does it matter if this term is 1 or 2 or 3 years too short or too long. Experience will refine our calculations. After all, the monopolies of the innovation patent would not cause the same interference with business generally that classical patent monopolies do.

BRIAN WRIGHT

Professor Wright has raised a number of queries which are very similar to some to which I have responded earlier, notably the "non-obviousness" criterion. In fact, this criterion interprets and enlarges the novelty concept, which at present only has meaning in the context of intellectual property.

In this context, the term "patentable" means nothing other than "appropriable", that is, subject to become intellectual property. So failure of this test does not at all mean that the invention or innovation "generates very little new knowledge not embodied in the patentable object". It simply means that one "expert skilled in the art" has taken a particular view, a view limited by his own personality, information and experience.

Professor Wright seems to share the misapprehension of some others that the innovation patent would deny patent protection for inventions. It does not. They are welcome as before, but they are granted patents (of innovation) only if (and when) reduced to commercial practice.

It is very important to grasp the difference in the ideological justification of monopoly under the two systems. As Wright points out, the (classical) patent is an "exchange of information for protection". The innovation patent, in contrast, is a reward for having reduced a concept to commercial practice.

This generates much more "information" than could ever be contained in the specification of a classical patent. Moreover, the actual innovation object also incorporates all the conceptual information which is included in such a specification. Normally, a specification includes nothing more than this conceptual information.

The adjustment for "capability", which Wright considers to be "misguided" is not "egalitarian" in its intent. It simply means that no monopoly shall be kept in force longer than is necessary for the owner to recover his costs and a decent profit, and this period is not affected by whether the patent owner is "efficient" or not.

In principle, the innovation patent system does nothing else than match the number of patents granted, with the number of innovative reductions to practice. In the classical patent system, there is no such matching; worse, the "non-exploited" patents even block other people from reducing concepts and ideas to practice. In which system, I ask, is the economists' "welfare function" better served?

On the question of market concentration, my only comment is that we should wait and see whether concentration is reduced more effectively by international competition than by an innovation patent system. In classical patents, foreigners can get patents which they do not work afterwards. Where, in such cases, is there any positive competition effect? With the innovation patent, foreigners can get an innovation patent if they "work" the innovation. Must this not result in more intense competition?

The innovation patent does indeed incorporate the patent of importation. Yet many countries (or groups of countries, such as the EEC) might well consider that their interests were better served through having a strong domestic industry than in relying upon innovations from abroad.

In the innovation patent system, too, the kind of objects patentable in different countries, varies according to their stage of development. Its criterion of "no prior public use in the particular country" will be recalled.

This means that many small firms can provide very simple technologies in or to an under-developed country. A "high-tech" firm will not be able to be of comparable service to such an economy, because its products are too advanced.

Incentives in the form of temporary monopolies should always be matched to the need and capacity for them in each individual case. It is precisely this type of matching that the innovation patent is able to provide.

GENERAL OBSERVATIONS

As the reader of the "critical" Chapters will have noted, the evaluation of the two proposals advanced is partly negative, partly sceptical, partly cautious, and, here and there, positive. Most of the critics have recognised the originality of the concepts, but all have expressed doubts about the workability of the models and about their chances of becoming accepted.

I am not altogether surprised, because intellectual and industrial property, especially patent protection, must be considered as part of a specific type of profane religion, sanctified by long tradition. It is therefore not easy to free one's spirit from deeply absorbed concepts and convictions, and have a new or unusual look at the role of the patent system in a fresh or original way.

Many of the comments, especially the more negative ones, derive from misunderstandings of the proposals or from that psychological phenomenon which is typified by projecting traditional logic and meaning on to elements which, although they may be old, are newly arranged and interlinked, as is the case here.

An additional factor is that patent protection is differently perceived and construed in the various countries of the critics, which may explain something which has surprised me: I would have expected the critics to devote much more attention to assessing the possible economic impact of the proposals, on the theoretical assumption that they *would* work.

It is quite clear that such an assumption can be contested or doubted relatively easily, and I admit that there is hardly an opposing argument in the criticisms expressed – even as to the usefulness of the whole patent system as such – that I myself would not have raised, as a critic.

But I would still have liked to hear something about questions such as the likely effect of the proposals upon competition, on industrial development, on international trade, on the relations between enterprises, and between enterprises and inventors, on cartel and unfair competition law, on acceptance of the patent system in society, and so forth, especially in comparison to the existing patent protection concept.

Because, if in these respects, the proposals could not bring about a better and measurable contribution to national prosperity, it would hardly be necessary to discuss their practicality at any great length. Starting the discussion from this end, however, involves speculation. We have actual experience of the classical patent system; but the possible performance of the systems proposed can only be grasped through imagination.

The basic question, of course, is: Who really wants reform of the intellectual property laws?

Could it already be too late to eliminate the inadequacies, incompatabilities and failures of the present patent law and patent system, not to speak of its complexity, cost and legal insecurity?

These disadvantages, of course, are completely hidden behind the healthy exterior which it exhibits to the non-initiated public.

My hope is that it is not yet too late for reform, and that even if patents degenerate so far as to become only quaint insurance policies for those who can afford to pay the premium, their place will be taken by something very much better.

It is in this spirit that I have proposed the innovation patent system.

CHAPTER XIII

Response by W. Kingston

When the distinguished commentators were invited to express their views on the two proposals for direct protection of innovation, it was insisted to them that they should not hold back in any way from adverse comment; it was important that wherever they saw a weakness in the proposals, they should point it out. Their work in response to this Brief, endorsed by their reputations as experts on Patents, guarantees at least one thing: It is highly unlikely that the proposals have any serious drawback that is not touched upon somewhere in the previous "critical" chapters.

This is of great value to anyone who has to evaluate the proposals, even more so to those who may be concerned in decisions on them. Nevertheless, however useful the critical chapters may be in putting the proposals into perspective, the ultimate object of dialectic is to arrive at synthesis. There remains the task, therefore, of seeing how well the proposals have withstood such detailed scrutiny, what constructive elements can be drawn from the experts' comments, and, finally, could a system of direct innovation protection be advocated confidently as a result, and if so, what would be its likely characteristics?

THE PRINCIPLE OF PATENTING AND THE "INFORMATION ECONOMY"

In his Chapter, Professor Piatier argued that the classical Patent system is not at home in the "information economy" that is developing; in theirs, Drs. Mandeville and Macdonald go further along the same path, rejecting the possibility outright that any application whatever of the Principle of Patenting could be beneficial. Mandeville and Macdonald are extremely articulate exponents of what may be called the Machlup tradition. In the previous Chapter, Dr. Kronz recalled that Machlup studied the Patent system for the

U.S. Senate, and concluded that if the system did not exist, it would be wrong
to establish it, but that since it did, it would be wrong to abolish it. After due
allowance is made for academic "hedging of bets" when faced with a problem
which has real-world implications, this amounts to a downright condemnation
of the classical Patent system.

Professor Wright also called attention to the way in which Machlup claimed
that inventions were "inelastic" in their supply, meaning that they arrived
whether or not there were financial incentives for them. This could well be
true, but it does not affect the arguments for direct protection of innovation,
since innovation, being a matter for *investors*, is very responsive to such
incentives, especially the all-important incentive of being able to appropriate
whatever rewards of the risky investment there may be.

According to the thinkers in the Machlup tradition, as Professor Tullock
records, and as Drs. Mandeville and Macdonald insist, intellectual property
can now do little to encourage innovation, and may even retard it. The reason
they give is that modern products and services contain so much more information than earlier ones. There need be no argument at least about the reality of
an historical change which can certainly be described in terms of the growth of
an information economy. This is easily seen, for example, in the contrast
between two prime movers, a modern gas turbine, and a Newcomen engine.
The contrast is to be observed, not only in the great difference in numbers of
components, but also in terms of the knowledge that has gone into the making
of each of these. Between the ironmongery of Newcomen's engine and the
rotor blades of a turbine, for example, there stands virtually the entire science
of metallurgy. Again, it is a very long time since it could first be said with
strict truth, that no aeroplane could lift the weight of the drawings that were
used in its making. Contrast this load (of information, note) with the lightness
of the information in the craft of the Wrights, of Bleriot or of Cody. Thirdly,
look at how much information is used in modern farming, in terms of animal
and plant breeding, fertilisers and pesticides, and equipment and techniques
for husbandry and harvesting, compared with the situation even a century ago.
And when to-day's fledgling doctor tends a diseased or injured patient, how
much more "information" is being brought to bear on the case than was at the
disposal even of a Semmelweis or an Addison.

To say, therefore, as Machlup was the first to do, that an "information
economy" is replacing the older type, is no more than saying that products
have more information in them than they used to have, and that all our
activities also have a higher information content. But this could have been said
with equal truth, at any time during the past two centuries. No doubt there
have been radical turning points, and it is likely that the coming of computers

is one of these, enabling us to handle greater volumes of information than ever before. Is it unique, however, as is implicit in Mandeville and Macdonald's evident enthusiasm for it? A good case could be made that in relation to the circumstances of the time, the advances in measurement, and in mechanical drawing and reproduction, of the early nineteenth century, were a similar quantum leap forward of enormous importance in terms of information handling; without them the shift to interchangeable parts would just not have been possible.

It is necessary to stress this point, because the protagonists of the Machlup tradition find it difficult to grasp that what we are dealing with to-day is not "a new heaven and a new earth", but much closer to being simply "more of the same". The result is that they fall into three main sets of errors, relating respectively to information transfer, to the size of the information economy, and, most important of all for the present study, to the validity of the Schumpeterian doctrine of the necessity of monopoly for economic innovation.

INFORMATION TRANSFER

In their concern to deny any need for formal means of protecting information, Mandeville and Macdonald lay a great deal of stress on innovation as a social process, in which firms obtain the bulk of the information that is important to them from outside,

"to be assembled in new patterns within the firm. Most information is not to be found lurking within the firm and very little is actually created there, no matter how strong its research and development department... Silicon Valley is outstanding in high technology industry because the participants in that industry acknowledge information as a fundamental resource and have adopted mechanisms to cater for its flow".

There is nothing particularly new about firms "obtaining the bulk of the information they require" from outside their own resources. The great German dyestuffs industry, for example, developed in the last decades of the nineteenth century on the basis of University work in theoretical chemistry; (1) all the electrical industries were built on the experimental work of "outsiders" like Faraday, Kelvin and Clerk Maxwell, Edison being the exception that proves the rule;(2) the many advances made by the German aircraft industry between the Wars, reflected its close co-operation with the University of Göttingen; State– and Foundation-supported medical research is used by pharmaceutical firms to an extent that makes it a major source of indirect subsidy to them; (3) the industries which provide inputs to the world's farms, have profited for well over a century from publicly-funded agricultural research programmes; and so on indefinitely. All sorts of industries, in fact, can be seen in their history to "acknowledge information as a fundamental resource, and have adopted mechanisms to cater for its flow".

Amongst these mechanisms is the firm's own R&D. It has long been recognised, in fact, that one of its most important functions is in enabling the information that comes from outside to be *absorbed*. Absorption takes place only as a result of the firm's own efforts to build upon whatever information is obtained from external sources, and will not take place at all without such efforts. But their nature, difficulty and cost will differ according to the stage of development of the industry.

Mandeville and Macdonald hold up Silicon Valley as the model for all industry in the future – especially in terms of the ease with which information is transferred and absorbed. This ease may be a function of an industry's age, more than anything else. In fact, the characteristics they list as being typical of modern high technology industries can be observed just as easily in the *early stages* of other industries. Piatier has referred specifically to the beginnings of automobiles and aviation, but Lancashire cotton, Swiss pharmaceuticals, German optical goods, even U.S. steelmaking, were all also characterized in their youth by, to use Mandeville and Macdonald's own words,

"an extremely rapid pace of technological change, intense competition, interaction among clustered firms, heavy and regular expenditure on innovative activity, and an important role for both small firms and individuals."

These authors also stress the way in which information travels "on the hoof", as they put it, in Silicon Valley, through the mobility of skilled and knowledgeable people. But there is nothing new about this either. Macdonald himself has chronicled elsewhere how much the great changes in agriculture in the eighteenth century, owed to the mobility of a corps of farm labourers who carried information with them. Arkwright's spinning frame arrived in the United States via the physical person of Samuel Slater, who memorised all the constructional details precisely before emigrating from England.(4) It is certainly true that all industries will become more like Silicon Valley in the sense that the products and processes of all industries will continue, as they have always done, to have constantly increasing amounts of information built into them. It is equally true that Silicon Valley will become more like other industries, as it, too, matures. Attention will be called below to the way in which this is already becoming evident in the way the information technology industries are seeking new ways to protect their innovations.

DECEPTIVE STATISTICS

Next, how big is the "information economy" and how much of it has anything to do with innovation? The enthusiasts naturally espouse estimates such as those which put 60% of all jobs in the U.S. into this sector. In pointing out that the "information sector" has been growing rapidly in all advanced economies, Mandeville and Macdonald suggest that it may now account for up to half the labour force in these. They implicitly link the growth of this sector with "the delicate working of the innovation process".

The truth, however, is that much of this growth has nothing to do with innovation at all. Parallel with the evolution of products and processes, which has been in the direction of ever-increasing information content, there has also been a massive process of bureaucratisation of society. Conventional statistics include all office workers in the information economy, and in doing so, they confirm Piatier's observation about the difficulties of interpreting statistics in relation to innovation. By these conventional criteria, for example, Marks and Spencers' ruthless war upon paperwork and record-keeping, with the clerical jobs that go with them, is stifling the growth of the information economy within that firm! Yet it cannot be denied that this is a core policy of one of the world's most efficient and innovative businesses in the distributional field.

Much of the growth in the information economy, as measured, therefore, involves the generation and shifting of masses of paper, and attendance at endless meetings, with only the most marginal generation of *new* information. There must even be the suspicion, therefore, that much of this growth as measured, may actually be anti-innovative. This suspicion is reinforced by the observation in the authoritative 1981 OECD report on the subject that "about two in every three people in the relevant industries are engaged in routine information-handling activities'; and by the identification of categories within the information economy by its authors as "directly unproductive, profit-seeking activities".

Having shown that the information economy has been evolving for two centuries, and that it is less advanced than conventional methods of measuring suggest, the next question to deal with is "What is the relationship between this evolution and protection for invention or innovation?"

Mandeville and Macdonald say bluntly that there is none, nor should there be any. They welcome the fact that the classical Patent system is too weak to provide any serious impediments to the beneficial flow of information between firms – "it works because it doesn't work", as they put it. This is hyperbole, but it is also inaccurate. The classical Patent system works very well indeed in respect of Chemical inventions, for reasons that were explained in Chapter I.

However, there can be no quarrel with Mandeville and Macdonald as to the poor performance of classical Patents in other fields, since this has been the stimulus for the present proposals. It is indeed the case that in these other fields, they do not provide any serious obstacle to the circulation of any new information that may be developed. And if new information did emerge without the need for means of appropriating the returns from risky investment in generating it, there could be no objection to the lack of any Specific means for bringing about such appropriability. Those who share Mandeville and Macdonald's attitude are presumably satisfied with the amount and types of new information being generated without any help from classical Patents. Tullock, in his Chapter, described this attitude as the "do nothing" approach to encouraging innovation.

INDUSTRY STRUCTURE

In fact, it means "Do not have a Patent system" rather than "do nothing". Those who advocate it are not proposing that the Company or Limited Liability legislation, or Trade Marks, should be abolished, and these also make innovation possible. There has to be *some* type of market power if investment is to be made in the generation of new information. As Nordhaus observed,
"External economies are an important aspect of the production of knowledge... If inventions are completely inappropriable, no profit-maximizing competitor will produce an invention, because increases in productivity would be instantaneously erased by a fall in price, and the firm would suffer losses to the extent of the research outlays."(5)
Complete freedom to imitate would totally destroy any incentive to invest in information production. This is true whether the information which results from the investment is expressed in a new drug, the precise shape, size and arrangement of components in a manufacturing process, an algorithm which can be the basis of a number of computer programs, or a developed and tested advertising campaign.

It remains axiomatic that money cannot be rationally invested at high risk except in the anticipation of correspondingly high reward. Schumpeter's insistence that this in turn demands monopoly is all the more relevant as the proportion of "information" in products and processes increases. Monopoly – market power, the power to keep others out of one's market, to prevent imitation – remains essential if new information is to be worth generating. Appropriability, as all of von Hippel's work demonstrates, is the key. At the limit, perfect power to imitate would ensure that there would be nothing worth imitating. Since this negates economic sense, various devices have been invented and developed to make imitation less easy, and so render investment in information production possible.

Professor Mansfield, in fact, has produced persuasive evidence that some important aspects of industry structure actually depend upon the effectiveness or otherwise of these devices:–

"Differences among industries in the technology transfer process (including transfers that are both voluntary and involuntary from the point of view of the innovator) may be able to explain much more of the interindustry variation in concentration levels than is generally recognised". (6)

In other words, where firms can keep the new information they generate to themselves, an industry tends to be dominated by a few large firms; where they cannot, it will be made up of many, smaller ones.

Mandeville and Macdonald's vision of the structure of industry in the "information economy" is the second of these, but they have not adverted to the fact that this demands measures that are deliberately designed to provide a counterbalance to the market power of either Capability or Persuasion, or both. If Specific market power is not available, or is inadequate to, or inappropriate for protecting investments to generate new information, this does not mean that there will be no such investments, nor that no new information will be generated. But it does mean that any investments will be made under the protection of Capability and/or Persuasive market power, and the new information will be different, because it will be of kinds that are appropriate to generation by these.

SIZE OF FIRM

This is where the question of firm size comes in. Both static and dynamic economies of scale constitute the barriers to entry to markets that are the essence of the Capability market power possessed by larger firms. In general, too, the larger the firm, the quicker it should be able to get an innovation to the market and so gain lead-time. Large firms can reap economies of scale in R & D and can also adopt a portfolio approach, at least to their run-of-the mill projects, in which one big success can pay for several failures. There are corresponding scale advantages in distribution and (especially) in advertising, that are the barriers in the case of Persuasive market power. As was stressed in Chapter IV, the absolute size of an advertising appropriation is a most important measure of its capacity to act as a barrier to entry. The very scale of the advertising expenditures of established brands can constitute a "virtual cartel" which is highly effective in deterring competition from newcomers.

Both these types of market power, therefore, inevitably tend to lead to large firms and to the kinds of innovation that are associated with large firms. Yet the price to be paid for size may be so high as to be fatal to innovative power, and the ongoing study of this needs to be taken into the evidence. The inevitable growth of the large firm's bureaucracy is antipathetic to individual creativeness. Its accountants ensure that no irrational decisions can be made – yet investment to innovate, being investment under uncertainty, *can never be fully rational*. The pressures for choosing the "safer" projects are often overwhelming.

Yet, great as these disadvantages are, they are probably unimportant in reducing the inclination of large firms to innovate, compared with one other factor: That being what is known as a "fast second", an imitator who deliberately lets others do the R & D and then uses his large resources to enter the market with a strongly-promoted imitation as early as possible, almost always pays better than being an innovator. Some statistics on this will be discussed below.

To the extent that large firms have a lesser tendency to innovate than smaller ones, that tendency has to be debited against the account of the types of market power that underwrite them. Not alone is Specific market power needed to ensure that there will be as much innovation as possible, it is also essential if the innovatory potential of small- and medium-sized firms and of individual creativity, is to be tapped.

Paradoxically, it is only the type of innovation protection which they attack so vehemently, especially in the direct forms proposed above, that could provide the economic conditions – and the prolific innovation – that Mandeville and Macdonald want to see.

INFORMATION TECHNOLOGY

It is important to stress this, because, as adverted to earlier, it is the information technology industry *in its extreme youth* which has caught the imaginations of Mandeville and Macdonald. It may very well be that when any industry is young, capability market power on its own is able to provide a healthy environment for small firms and for the creativity of individuals. It certainly cannot do so later on.

An indispensable characteristic of such an environment is that the stakes for entry are low, meaning that new firms can come into being easily, and that individuals have little difficulty in moving from employment to start up their own firms. Digital Equipment, which is a paradigmatic example of the exciting early days of the computer revolution, was founded by an ex-employee of IBM, and is now worth hundreds of millions of dollars. It is a useful reminder of just how low the stakes were in those heady times, that Digital's starting equity was no more than $116,000. (7)

The ability of talented individuals to start their own firms is obviously greatest, not only when investment levels are low, but also when many of the characteristics of a service industry are present. There is a revealing analogy between those early days of information technology, and the advertising industry – Silicon Valley, in fact, looks extraordinarily like Madison Avenue. Using Mandeville and Macdonald's own descriptive phrases, each in its own way "urgently gathers vast quantities of information and explicitly recognises its dependence upon external information sources." "Personal information networks" are self-evidently essential to the advertising industry – where is the business lunch more important? As to information travelling "on the hoof" so that it can "provide full and timely input to a firm's innovation efforts", the mobility of creative people between advertising agencies is a by-word, and this mobility is associated with the development of new campaigns, which is one form innovation takes in the advertising context.

It is important to understand, however, that underlying the advertising industry is the legislative foundation of Trade Mark registration, and – to a lesser extent – Copyright, which together give a high degree of appropriability of rewards from the risky and large-scale investment that mass-market advertising requires. Monopoly in a Brand ensures that most returns from this investment can be captured by the investor in it, and will not be diverted to its imitators.

The monopolies which Trade Marks confer are clearly-defined, inexpensive and strong. They underwrite an immense amount of economic innovation (certainly "new things" in marketing are at least of the same order of economic magnitude as "new things" in technology to-day) and a corresponding amount of human creativity is provided with outlets. A huge number of small- and medium-sized firms in service industries (with which large firms do not mix) is kept in being by the Trade Mark monopolies (Brands) held by their clients. It is claimed for direct protection of innovation, that it would have similar beneficial results in other areas of business and in technology.

YOUTH AND AGE IN INDUSTRIES

There is no argument with Mandeville and Macdonald either, on the need to encourage the maximum number of independent points of action. The issue is whether or not the "do nothing" approach (which, as has been seen, does not mean "do nothing" at all, but "leave it to Capability and Persuasive market power") or the direct protection of innovation approach, is better. One weakness of "do nothing", is that it is impossible to see how products and processes involving *long-term* R&D (and thus with a high information content of a particular type) could come into being, if investors could not foresee an acceptable level of protection from copying, of any results there might be.

In the case of information technology, the market power of capability on its own was enough to deliver reward for investment while the stakes were low and the productive units small. For example, as long as the number of man-hours involved in developing a new computer program is counted only in hundreds, the industry can operate just as Mandeville and Macdonald describe it, with informal information networks, mobility of experts, frequent establishment of new firms, and *absence of intellectual property*.

It is a very different matter, however, once the relevant man-hours can run into millions, as they do now. The investment which this involves simply cannot be made without formal mechanisms to enable the rewards to be appropriated, and it is no coincidence that the growth of investment levels in the information technology field, has been matched by ever-increasing concern by firms in that field, to bring such mechanisms into being. Since so much of the criticism of the present proposals by Mandeville and Macdonald arises from their study of the microelectronic revolution, it is worth while dwelling on the story of formal protection means for computer programs. This illustrates clearly that, whilst the so-called "do nothing" approach to protection may work during an industry's youth, when the market power of capability alone suffices to give enough protection from imitation, this approach becomes progressively inadequate as the industry develops, investments become larger, and the life of products longer.

The problem about Patent protection of computer programs has been twofold. In the first place Patent Offices have been as unwilling to protect an algorithm – "a prescribed set of well-defined rules for solving a problem in a finite number of steps" – as to protect a scientific principle. Yet from the program writer's point of view, anything less has little value, since in that case all an imitator has to do is change the form of the program somewhat to avoid the protection. Secondly, the time it takes to grant a Patent is a most serious drawback when the life of programs is short.

Similarly, Copyright has little real protection to offer. Only the program *format* is covered, so once again an imitator can use the valuable algorithm freely, just by re-writing the format differently. Trade Secret protection, too, is inadequate, although at least it does operate from the moment the program needs to be protected. IBM proposed a Registration system with some Trade Secret elements in 1969, and this idea received support from the World Intellectual Property Organisation.

NEW LEGISLATION

Nothing demonstrates the inadequacy of Capability market power on its own for protection of major investments in the information technology area, better than the U.S. Apple-Franklin case, which was finally settled on Appeal in 1983. The question was whether Copyright protection applied to a program incorporated in a chip, and what was claimed to be at stake was Apple's very survival in the face of a flood of "lookalikes", imported from the Pacific Basin. Settlement in Apple's favour led to the passing of the Semi-conductor Chip Protection Act in 1984. This gives a ten-year term of protection and makes registration mandatory.

In relation to criticism of the present proposals, however, one of the most interesting aspects of it is its relation to the traditional Copyright law, which is Chapters 1-8 of Title 17 of the U.S. Code. The U.S. is a member of the Universal Copyright Convention, so if Copyright on Chips was simply added to this, under the doctrine of "national treatment" it would automatically apply to foreigners also. By making it a new and separate Chapter 9 of the Title 17, this Convention requirement is evaded, and Chip protection is only going to be given to nationals of countries that genuinely reciprocate. The target of this provision is explicitly the Japanese, a point which is called to the attention of those who may consider the argument in Chapter IV, about the need for new forms of innovation protection to counter Japan's Capability market power, to be exaggerated.

Further, the existence of the new Act, brought about by pressure from the Computer industry itself, must move that industry another step away from the "personal information networks, highly mobile personnel, information as a common resource" pattern so highly approved of by Mandeville and Macdonald. The change is simply a function of scale of investment and the time needed for its recovery. As these increase, there is progressively more need for specific legal instruments for innovation protection.

A DYNAMIC INFORMATION ECONOMY

The Principle of Patenting offers the possibility of an entire economic environment that is highly conducive to innovation, with all the features which

Mandeville and Macdonald observe with approval in the information economy, and more. This is difficult for disciples of Machlup to grasp, because they naturally share his awareness of how poorly the classical Patent system performs, and think that this is the only expression which it is possible for the Principle behind it to have. But the proposals of Chapters II and III represent two overlapping ways of putting this Principle into effect, which have none of the disadvantages of classical Patents.

It is indeed true that these Patents have become little more than a reinforcement of the Capability market power of the largest firms, but this need not have been allowed to happen. Either the Innovation Patent or the Innovation Warrant would grant genuine, incontestable monopolies which would enable innovation to become a commonplace activity of smaller firms. This cannot happen at present, because a classical Patent on its own offers a poor basis for investment, and firms of this type do not possess enough of the other types of market power.

A particular feature of the Innovation Warrant, is that it contains provisions for ending the advantage possessed by large firms in virtue of their resources for litigation. As will be seen below, an adaptation of some ideas of Professor Tullock's offers an effective means of putting this provision into practice, and this is equally applicable to the Innovation Patent. The result, it will be evident, could be a major factor in making investment in innovation a rational activity for smaller firms for the first time. Since there are so many of these, opportunities for mobility of skills and creative energy must be multiplied compared with present levels. Innovation Patents and Warrants would also frequently provide a valid basis for starting new firms, thus making the general industrial environment very much more like the dynamic information economy that Mandeville and Macdonald want to see.

COSTS OF PROTECTION

It is perfectly true, as these authors point out, that there are costs involved in any system of protecting innovation and that no system can "convey only total social benefit". Tullock has adverted to the same point. But what matters is the ratio between the benefits and the costs, and it may fairly be argued against Mandeville and Macdonald that they do not advert to the costs which result from Capability and Persuasive market power. It is claimed that the variable term of monopoly, which is common to both the Innovation Patent and the Innovation Warrant, would be a powerful factor in achieving a high benefit/cost ratio in the case of direct protection of innovation.

The Innovation Patent sets out to achieve the ideal, which is matching of length of monopoly exactly to what is required to bring forth the innovation in each individual case. Such matching would indeed cut "restrictions on use, higher prices and inefficient methods of production owing to insulation from competitive pressures", about which Mandeville and Macdonald are apprehensive, to the absolute minimum. The "categories" proposed for the Innovation Warrant could not achieve the same perfection of matching in each individual case, but should nevertheless deliver high ratios of benefit to cost.

Either approach, or an amalgam of both, must contribute strongly towards providing the "clusters of competing firms" which Mandeville and Macdonald rightly consider to be the ideal condition for rapid innovation. Not alone would innovation be possible for large numbers of firms which could not otherwise afford to contemplate it, but the variable term would throw any new technology or newly identified market open to all-comers for development at the earliest possible moment.

It is realistic to accept, however, that there could be cases where the system fails, in the sense of giving a longer monopoly than is consistent with the most rapid progress possible, because the resources being brought to bear are those of only a single firm, rather than of a multiplicity of firms. The problem here is to find a way of opening up the innovator's territory, without reducing his reward, and a suggestion for dealing with it is advanced in the section on the term of protection, below.

UNEMPLOYMENT, PROTECTION AND NATIONALISM

Professor Tullock, in referring to the fact that innovation is not a universally approved activity, observed that this is largely due to ignorance on the part of those displaced from work, of the reality that human wants are limitless, so that there will be compensating employment in new industries. Professor Silberston takes the same line, although he accepts that "there may be severe transitional difficulties".

Both reflect the influence on economists of the ideal of the perfectly competitive market. This ideal, however, never did exist and never could exist in the real world, and is in any event incompatible with economic innovation. The modern world has been accurately described as a "world of monopolies".(8) In economic terms, it is structured in terms of market power, and the distribution of this power as between countries has a major bearing, not alone upon the location of innovative activity, but also of employment.

Economic change of all kinds, which is instantaneous in the ideal (but impossible) model, becomes "change with a lag" once the rigidities of real life are introduced. This lag is especially observable in respect of employment. If new technology makes a particular group of workers redundant, Tullock claims, it also makes whatever they had been making, cheaper. Consequently, more money is available to pay for alternative (sometimes quite new) goods, and demand for labour in providing these, increases.

This ignores two important points. The first is that once market power becomes a reality of economic life, competition is no longer only in terms of price, but progressively relies on factors such as advertising and service. Where this type of competition is widespread, the innovations which bring about unemployment will not necessarily be associated with price *decreases* in respect of the goods to which they apply. Instead, the economic result of the innovations is likely to be increased rent to the firms in which they take place, not more widely dispersed purchasing power amongst consumers generally.

Further, even where innovation does lead to lower prices, the pressure of sales promotion will tend to direct the resulting "freed" purchasing power towards products manufactured by capital-intensive industries. Marginal increases in the sales of these, may simply improve their productivity through raised plant utilisation, without any additional requirement for workers. Thus, even without taking account of rigidities in the labour market itself, the structure of modern industry is not apt for absorbing labour displaced by innovation.

Labour market rigidities are the mirror image of those which have become typical of capital, and are most intelligible as derived market power. They evidently add greatly to the difficulties of re-deploying labour which innovation has made redundant. But in so far as they are "derived" they must automatically be ameliorated if rigidities in terms of capital are lessened. It is innovation that undermines established capital structures, as Schumpeter's famous passage about "creative destruction", already quoted, reminds us. Consequently, if Direct Protection is the cause of more innovation, it must contribute to breaking down the rigidities of capital, which in turn must reduce rigidity in the labour market, and this cannot fail to reduce unemployment.

There is also a geographical aspect to the employment problem. It is no coincidence that two areas in the world that have had massive employment growth, the U.S. and Japan, are those that excel in economic innovation, although of different types. Europe, in contrast, has only increased employment by about 2% in a generation, and also lags badly in terms of innovation. For European countries, therefore, it is of particular importance to take advantage of the contribution which Direct Protection of Innovation could make, on both counts.

Tullock's view that specialised skills have now been largely replaced by "a skill in adapting to new situations" also seems excessively sanguine. It does not seem to apply, for example, to very large numbers indeed of workers in the U.S. automobile industry, in British shipyards, and in steel works throughout both the U.S. and the E.E.C., all of which have been affected by Japanese productivity, itself the result of innovation on a massive scale. The "severe transitional problems" referred to by Silberston may have a time-scale as extended as Keynes's famous "long run". In that case their "transitional" aspect is sadly irrelevant to the present generation and probably also to the next.

The Chairman of Phillips NV. recently delivered an address on the theme that present trends are leading to a situation where "the United States will feed the world, Japan will provide it with products, and Europe will be their playground". Dr. Dekker is indeed in a position which gives him exceptional access to information on the competitive problems faced by European industry. But what if Europeans do not wish to be transferred into the only alternative employments which current patterns of innovation appear to be offering them? It is hardly surprising if a group of people whose history includes so much creativeness and originality, not least in technology, should reject the idea that their future is to be seen only in terms of international tourism.

FREE TRADE AND ECONOMIC POWER

In considering what the appropriate response should be, it is important to be clear-sighted about the realities of free trade and protection. World-wide free trade has always been advocated by whatever country is dominant in terms of the market power of capability at any particular time. In the nineteenth century it was Great Britain; in the post-war era it was the United States. The General Agreement on Tariffs and Trade, which the U. S. took the initiative in founding in 1947, had for its objective the removal of barriers to the movement internationally of the new types of goods in whose production U.S. firms excelled. The "Kennedy Round" of further tariff reductions reflected the same dominance of U.S. capability before the Japanese thrust began to be felt in earnest – at a time when Japanese cars were still a joke to Americans, it will be recalled.

To-day, with G.A.T.T. manifestly in tatters, it is the Japanese who lead in the market power of capability and it is they, not the Americans, who are pressing for world-wide Free Trade. They face restrictions of all kinds, from so-called Voluntary Import Limitations to "Poitiers-type" Customs clearance delays. These are so widespread that even Japan's own productivity growth figures are distorted. The falling-off in the latter in recent years (into which some Western economists have actually read hope of slackening power to threaten our industries) has only taken place because Japanese firms have been prevented from continuing to penetrate Western markets at an ever-increasing rate, by the erection of non-tariff barriers to international trade.(9)

One side of this issue is stated succinctly by Silberston:–

"The consumer (as opposed to the home producer) might benefit greatly from low-cost imports. Is the encouragement of inefficient home industry to be given priority over the welfare of the consumer?"

But there is another side which can be put as follows:–

"Is consumer welfare to be thought of only in terms of what, and how much, the consumer can obtain by his purchasing power, without reference to how that purchasing power has come to be in his hands?"

The Free Trade versus Protection argument cannot be carried on only in economic terms: It involves strong cultural and political elements as well. The argument of Chapter IV is that if harm is done by protection, that harm is minimised by restricting any protection given, to innovation. Similarly, if protection is capable of producing beneficial results, these are most likely to be obtained if it is limited to innovation.

In the actual situation faced by politicians, protection of innovation is far better than erection of non-tariff or other barriers to international trade, because it relates to the industries of the future, not to those of the past. It is the very opposite of a "siege economy." Innovation has to be protected anyway, if investment in it is to be a rational activity, and this is true irrespective of whether it is taking place in the most efficient firm in the world, or in an inefficient one.

MEANING OF EFFICIENCY

Indeed, it is an important question whether the usual criteria of efficiency have any relevance at all to the activity of innovation. Consider the Float Glass example raised by both Silberston and Tullock. Innovating this was a major struggle, with many setbacks, for Pilkington's. Yet during this period (which involved losses of a relative magnitude which they hold could never have been accepted if the firm was a public one) they were the most "efficient" float glass manufacturer in the world, since they were the only one. It was only later, when their fully developed technology was licensed to firms in a number of countries, and had effectively displaced the traditional grinding techniques, that measurements of efficiency in the usual sense could begin to have meaning.

Note, too, that according to these measurements, Pilkington's production of glass by the float method throughout the long innovatory period was highly inefficient compared with its production by the traditional method; worse, during this same period, bearing the costs of the float glass development would have made the entire Pilkington firm look as if it was growing progressively more inefficient, by any or all of the normal criteria used by accountants and economists. Such criteria favour firms at the end of one experience curve, and penalise those at the beginning of the next one. If, therefore, the concept of efficiency/inefficiency loses its meaning when applied to innovation, can any arrangements which are strictly limited to innovation protection be validly criticised from the standpoint of "efficiency"?

NOVELTY AND GEOGRAPHY

It will be recalled that both forms of direct innovation protection proposed, allow for the possibility of novelty being less than universal. The criterion is "not already available in the ordinary course of trade". This could mean, not available "anywhere in the world"; "within the E.E.C."; "within a particular country"; or – in the extreme case – "from indigeneous national production". The latter, it was made clear, is envisaged as usable only by Third World countries, and it is accepted that it is open to the charge of being a regime of national protection, albeit with a strict time limit. It may be observed in passing that any use of the device of protection by such countries to develop an industrial base, would be no more than following in the footsteps of the United States and Japan, both of which used this means, with conspicuous success, to get their start in industrial life.

But the argument advanced above, that "efficiency" criteria are not applicable to innovation, renders it easy to defend the use of a restricted novelty criterion by Third World countries. Innovation protection is always protection of new information; in the circumstances of poor countries the new information that matters above all, relates to how to carry out productive activities successfully, in other words, the process of "learning by doing". The biggest gap in the armoury of Third World development, is the lack of experience of achievement, simple step by laborious step. Nothing can be more important than the protection of the information which is appropriate to this experience. For Third World countries, innovation is the application of *existing* technology to *local* conditions. Classical patents cannot protect this activity, but the Innovation Patent and Warrant systems are perfectly suited to it.

INTERNATIONAL PATENT FLOWS

New quantification of this reality has recently appeared and confirms the argument in Chapter 1 about the imbalance of advantage in membership of the Paris Convention. In 1980, the U.S. and Japan granted an average of 0.43 Patents to foreigners for every Patent they obtained abroad (and for the U.S., one in four of all Patents of the former type was to a Japanese). The comparable figure for a group of 16 other OECD countries was 3.3 and for 13 "semi-industrialized" countries, 4.5. As for the Third World countries, the most advanced of them in this respect, India, gave 2000 Patents to foreigners in 1980, but Indian firms obtained as few as 57 Patents abroad. The actual economic imbalance must be far greater than these figures show, since they are not weighted for quality in any way. India's "foreign" 57 Patents are unlikely to match the technology of the Patents obtained in India by firms from other countries. The reasoned conclusion of the author of this study, Professor Robert Evenson, deserves to be quoted in full:–

"Implicitly, these international conventions seek to provide global (or as much of the globe as possible) property rights to inventors in a particular country. This may be a perfectly reasonable trade agreement between certain countries (e.g. EEC countries). We have observed in this paper, however, that trade in intellectual property is a very unequal trade, with developing countries having a strong competitive disadvantage in supplying intellectual property to developed country markets. Their inventors do not have the economic laboratories and other resources to enable them to be competitive. Ironically, nations do not recognize global rights in non-intellectual property and regularly intervene in commodity and capital trade markets to achieve nationalistic goals. With few exceptions, these same nations have joined international conventions freely granting intellectual property rights to citizens of other countries. By doing so they have gained some advantages in bargaining with multinational firms and in some forms of technology purchase. But unless the cost of "pirating" inventions is very high they have paid more than necessary for technology purchased from abroad.

However, the most serious impact of membership in international conventions may well be that it restricts the flexibility of many countries to design legal systems tailored to their comparative advantage, particularly regarding adaptive invention and the encouragement of indigenous secondary technology core development."(10)

EUROPEAN RESPONSE

The Innovation Warrant and the Innovation Patent both represent attempts to design legal systems appropriate to the needs of countries at several levels of development. The criterion of novelty which involves indigenous production, if used at all, should be restricted to Third World countries. For the OECD group, in any event, there can be no question of adopting the most restrictive definition. The "new in the whole world" criterion would deflect all criticism on the ground of protecting inefficiency, but it would not be difficult to defend "regional" and even "national" limitations. "Not available in the ordinary course of trade within the E.E.C.", for example, would be a perfectly sensible and easily administered novelty criterion.

Professor Silberston expresses reservations about the way in which the proposals, specifically the Innovation Warrant, "bring a much more nationalistic flavour to the protection of intellectual property than has hitherto been the case." For Professor Tullock also, "the geographical provisions on the innovation protection are perverse". This is in spite of the fact that Tullock accepts that "for almost all small countries, they are better off if they do not honour foreign patents", and in fact he demonstrates theoretically how this must be so.

Tullock further observes that "the geographic limits of the innovation patent or warrant offer an almost perfect excuse" for evasion of obligations under the Paris Convention, and he sees no point in putting such a temptation before smaller countries which, as he says, "have strong motives to try to evade the rules".

In this he is not strictly correct. The Innovation Patent has no geographic limits, in that it can be granted for imported products. It therefore offers no greater temptation to smaller countries to evade the rules than the classical Patent system does. Tullock's criticisms therefore only apply to the Innovation Warrant. Yet even the Warrant is compatible with membership of the Paris Convention as long as the principle of "national treatment" (equality before the law, for nationals of all Convention member-states) remains intact. The Warrant is open to natives and foreigners on equal terms, but does require that the investment in the innovation should be made locally.

This causes Tullock to argue that "Entrepreneurs would put up factories in areas and places where they were not actually economically justified on the grounds that with a monopoly they would be justified". But this overlooks the reality that the present geographical distribution of factories and employment already largely reflects monopolies or quasi-monopolies, although of a different kind to Patents and Warrants. This distribution is certainly a reflection of the world-wide balance of the market power which actually underwrites innovation.

Apart from the distortions introduced by Law, the "free" market, of course, is also subjected to direct intervention of all kinds by governments. There seems little reason to doubt, for example, that if it were not for such measures, Japanese firms would prefer not to manufacture or even assemble products in advanced countries. Also, given free trade, U.S. firms could be expected to close down many of their offshore manufacturing facilities. Direct protection of innovation would do no more than redress what is, in any event, a highly artifical set of arrangements. However, it would do so in a way that introduces more rationality into Governmental interventions. This way also corresponds to the individualistic tradition of the West, and offers opportunities that are particularly attractive for European industry.

It is time for countries that are economically weaker than the U.S. and Japan to take account of this reality in their Intellectual Property policies. The U.S. is dominant in terms of Persuasive market power and is likely to remain so, given the way in which the size and nature of its home market supports the development of branded goods, advertising and material for the international mass media. Japan's comparative advantage in terms of incremental innovation, is equally unassailable. Europe's response can only be in terms of developing new types of Specific market power with which to underwrite innovation, since its own laws remain under its own control.

SILBERSTON'S OBJECTIONS

Professor Silberston doubts whether processes would be protected effectively by the proposals. The Innovation Patent has specific provisions for protecting processes through their products. Nevertheless, Silberston's doubts are reinforced by the empirical research of the Yale survey (to be referred to in detail below) which shows clearly that the existing Patent system is considered by R&D Managers to be even less effective in protecting processes than products. It is therefore accepted that the Innovation Warrant, too, should explicitly apply to processes in exactly the same way as to products.

He also feels that "the proposed co-existence of the classical patent system, on the one hand, with the Innovation Patent or Warrant (or some amalgam of the two) on the other, seems to be a great obstacle to the proposals." He thinks that "where legal monopolies are being granted by the State, only confusion can be caused by two parallel systems," and he is sceptical as to whether the new proposals can simply be added to the existing Patent system.

It should first be stressed that such an addition is not included in either proposal. The Innovation Office would not be the Patent Office. Secondly, States already run several monopoly systems in parallel. Copyright and Design Registration are very close (indeed, they virtually overlap in countries where copying of three-dimensional objects infringes copyright in the drawing from which the object was made). Trade Mark protection, too, is only marginally distinguishable from Copyright protection in the now commercially important area of "character merchandising." No substantial difficulties appear to have arisen as between such parallel monopoly systems, nor need they be expected between indirect and direct protection of innovation.

The principal arrangements proposed in Chapter II in respect of the Innovation Patent's relationship with classical Patents, would also be adopted for the Innovation Warrant. These are:–
1. Where an innovator has actually begun his work before the (classical) Patentee files his application, he will be allowed to continue, following the traditional Patent doctrine of "Prior Internal Use".
2. If a classical Patent is not exploited, after one year the Patentee would have to issue a disclaimer of rights in respect of any overlapping part of an innovation object. This is a variant of the usual loss of Patent rights after three years, for non-use.

DE JONG DOUBTS

Professor de Jong's arguments against weighting the monopoly term according to "innovative capacity" (for the Innovation Patent) or to categories of firm size (for the Innovation Warrant) will be discussed in detail below. His recommendations on this point will be accepted.

He also doubts the capacity of an Innovation Office to handle an examination for innovative novelty. It is just as difficult, he claims, to establish whether a product or a process is new, as whether an idea is new, if not indeed more so.

Reflection on this point suggests the need to distinguish between capacity to recognise that something is new, and capacity to grasp its meaning. As was stressed earlier, the meaning only becomes evident retrospectively, but the fact that something is new, that is, different from the current state of the Art, may be immediately obvious.

de Jong is therefore right to doubt the ability of the staff of a public institution to assess innovations. But that is not what is expected of them. All they have to be able to do, is recognise that something is marginally different, radically different, or different in some intermediate sense, from what is already on the market. They are not called upon to make any judgment as to whether it is, or may eventually be important.

The fact that bureaucrats are notoriously incapable of acting in the face of risk, much less uncertainty, does not mean that they cannot recognise these when they see them. The experts who, in case after case, failed to recognise that some ideas submitted to them were going to be the basis of whole new industries, were in no doubt as to whether or not the suggestion was new; the problem was that it was *too* new (i.e., it involved too much uncertainty) for them to recommend investment by their firms.

Incremental innovation is the easiest for all concerned to recognise, since by definition it emerges naturally out of the existing technology; radical innovation is equally discernible, by the very absence of such technological pointers, as well as of opportunities for reliable market research. Indeed, it might even be definable by the very lack of factors which would enable risk to be calculated, so that any decision in respect of it would have to be taken under uncertainty. An innovation of the intermediate level of novelty might be distinguishable by the relevance to it of the new market research techniques which study the *users* of innovations with special care.

Thus, at least for the three categories to which variable terms of monopoly might be related at the outset, on de Jong's own recommendation, there should be no insuperable difficulty in deciding what is "new." The resources for research which the Innovation Office would have at its disposal could also reasonably be expected to improve the Office's capacity in this respect over time.

Professor de Jong also "would not be in favour of extending the area of protection too widely beyond technological innovation." There is not the slightest difficulty about accepting this recommendation. Direct protection of innovation would obviously be tried out first in the most appropriate fields, and only when it is proved to be successful and when experience has been obtained in actually working it, would there be any question of extending its scope.

Engineering and Electrical innovations have claims to be the first to get Direct Protection, since these have been so badly served by the classical Patent system. It would be desirable, however, that an Innovation Office should be empowered by any legislation setting up a system, to broaden its range beyond whatever it starts with, at its own discretion.

There could also be a case for restricting Innovation Patents and Warrants at the outset to applicants in a particular region (although the monopolies granted, of course, would apply to the whole country). Hitherto, efforts to correct structural and regional imbalances have relied heavily on financial transfers, and it cannot be claimed that they have given good value for money. Limiting the geographical area within which an Innovation Patent or Warrant could be obtained to wherever regional aid is currently available, would be a means of measuring what Law could do, where Intervention has had so little success. The results of any such restrictions in national schemes would be of particular interest to the European Commission, which, since 1972, has been grappling with problems arising from the objective of reducing the differences between the various regions, which is a feature of the Treaty of Rome.

COPYING WITH IMPUNITY

The insistence on the growing importance of information in the economy, in the contributions of Piatier, Mandeville and Macdonald, has been of outstanding value for the present study. It adds a whole dimension to our understanding of why the classical Patent system no longer works well, and also shows how the proposals for direct protection of innovation could be improved.

The measure of both classical Patents and the new systems proposed, is the extent to which firms or individuals are free to make use of information which they have done nothing to bring into being. The question of copying is critical and the evidence that the classical patent system's performance in this respect is altogether inadequate, continues to mount up.

In a perfect system, the time taken by a competitor to put a rival product on the market should at least equal the full term of the Patent, and the cost of doing so in any lesser time would be infinite, because impossible. In his contribution, M. Bouju called attention to how the work of Professor Mansfield of the University of Philadelphia shows that the reality is very far from this. Consider first, Mansfield's results in respect of electronic and mechanical inventions, each of which cost less than $1 million to innovate. In *not even half* of these cases (48%) did the time taken to imitate exceed what it had earlier taken to innovate.(11)

Worse, the *cost* to the imitator only actually exceeded that to the innovator in 11% of cases. In respect of inventions whose innovation had cost over $1 m., the figures are 31% and 22% respectively. It is reasonable to assume that the more costly innovations were undertaken by the larger firms, so that the somewhat better results for these, also reflect a substantial degree of capability and/or persuasive market power. In any event, the ineffectiveness of Patents is clear, as Bouju noted.

YALE STUDY

Mansfield's figures have now been confirmed by the very large and thorough Yale study of the methods of appropriating the results of investment in innovation actually used by R&D Managers. Bearing in mind that the primary function of a Patent system is to prevent imitation, and that a perfect system would do so completely for the lifetime of a Patent (17 years in the U.S.) it is a devastating comment upon the actual performance of the system, that effective duplication of a typical Patented product was considered to be possible in *less than 3 years* by 90% of respondents. And *only 8%* of the responses from these experienced R&D managers were to the effect that imitation in such cases would cost the imitator more than it had the innovator! Even when an imitator does have to pay some premium for not having to carry the uncertainty which the innovator has faced, it is not high:–

"We find that patents raise imitation costs by about 40 percentage points for both major and typical new drugs, by about 30 percentage points for major new chemical products and by 25 percentage points for typical chemical products. In electronics, there is some discrepancy in the results for semiconductors, computers and communications equipment, but the range is 7 to 15 percentage points for major products and 7 to 10 percentage points for typical products". (12)

All the above figures, it should be remembered, relate to one of the best Patent systems in the world. If the U.S. system is so ineffective in delivering the main thing which is expected of it, which is freedom from copying, what must be the situation in those many countries whose Patent systems are less rigorous?

Further, an innovator, as some of the work of modern economic writers has been concerned to demonstrate, is operating under uncertainty. What this means is that the underlying stochastic processes are not regular enough to allow him to estimate the probability alternatives, and act accordingly. He is under the great disability for a business man, therefore, of not being able to apply a portfolio approach to his innovatory projects. In respect of any single project, he is highly unsure that he will achieve technical success, and even if he does, whether it will be in the form which the market wants. (13)

The imitator, in contrast, may be said to be investing his money on the basis of risk rather than uncertainty, and can even obtain relief from risk by diversifying amongst projects. He knows the answer to the all-important market question, and has a quite different – and much easier – technical problem to that which faced the innovator; once new ground has been successfully broken, it is obviously far less difficult to achieve the same result in a way that avoids the Patent system's restrictions.

For business men and investors the system stands or falls by how well it removes one part of the uncertainty that attaches to investment in innovation. When it is considered how much uncertainty is removed between the two stages, there can be little doubt of the strength of the rational case for investment in imitation rather than innovation. If 60% of all products that are supposed to be protected by Patents are successfully imitated within four years, as Mansfield found, investment in innovation rather than imitation on the basis of Patent protection alone, *cannot be a rational exercise*. Both the Mansfield and Yale figures confirm the claim in Chapter I that most of the investment that is actually made in innovation – assuming that this is at all rational – must be being made on the basis of types of market power other than the Specific market power of the Patent system.

The two proposals for direct protection of innovation are attempts to change this situation, and to offer genuine incentives for investment under uncertainty. These are real monopolies, rather than the spurious ones of the classical Patent system.

In turn, this means facing the problem of copying squarely. Most existing systems express the prevention of copying of an invention through use of equivalents in the technical sense. It is often possible, however, to obtain the same or a very similar result from the customer's point of view, which is what matters for sales and profits, by means that cannot be considered to be technically equivalent. Some Patent administrations, notably that of West Germany, outlaw "legal equivalence" which covers this type of imitation also, and consequently their Patents confer a higher degree of "appropriability". The Innovation Patent proposal, it should be noted, uses "technical" but not "legal" equivalence. The comments which follow, therefore, relate primarily to the Innovation Warrant, which introduces the doctrine of "commercial equivalence".

According to this, infringement is an act which uses any new information generated by the Warrant-holder, to his detriment during the monopoly period. This meaning of infringement is in sharp contrast to those of classical Patents. As the Mansfield and Yale figures prove, the limits of the monopoly power when the criterion is "technical" or "legal" equivalence, increase the cost of copying somewhat, but by no means prevent it. The doctrine of "commercial equivalence", coupled with incontestability, is intended to prevent it, positively, clearly and completely, until the end of the monopoly term.

INADEQUACY OF "CLAIMS"

This doctrine is perfectly in line with the growth of the information economy. To the extent that all products, processes and services contain more information, it becomes progressively impossible to delimit what is "new" in them, in terms of the "Claims" of the classical Patent system. That system has been criticised in the past for "attempting to break up the flow of creative energy into discrete sections". It is now increasingly vulnerable to the charge that it similarly attempts to break up the flow of *information*. Indeed, there is a sense in which the modern evolution of Patents towards protecting invention rather than innovation, reflects their decreasing ability to force the information content of innovations into the strait-jacket of Claims.

Piatier's analogy of the "flow" of innovatory information is helpful here. Just as the further downstream, the more tributaries have contributed to the composition of a river's water, so in innovation, the information content "downstream" is more complex than it is "upstream". Patent "Claims" can only cope with a limited amount of complexity. As innovations become more complex through containing increased amounts of information, the gap between what can be captured in "Claim" terms and the full range of the novelty widens.

Recognition of this, combined with inability to see how adherence to "claims" could be reconciled with growing information content in products and services, therefore had only one possible outcome: that the classical Patent system would have to move "upstream" from innovations, to make abstract "inventions", with their very much less complex information content, its subject-matter.

Mandeville and Macdonald are perfectly right to stress how modern, innovatory firms are debtors for information to a whole range of external sources, and that very little information is actually generated by the firm itself. Yet that small contribution by the firm represents the essential difference between that information being used or not, in some way that is of practical benefit to the public. In any actual innovation, many pieces of information have been generated by agents other than the innovator. They are in the public domain, and an innovator uses them in combination for the first time in a product or process. Bearing in mind that all this information may have had a number of practical uses before, how is it possible to identify what is the precise element of novelty which the innovator has added? And yet it cannot be denied that he has added something, since a product or process now exists, that is different from what existed before.

USE OF "HINDSIGHT"

There is one way out of the difficulty. It is of the very nature of all discovery that the nature and range of what has been discovered is not known, nor, indeed, is it known how the new information relates to pre-existing knowledge. Not alone Whittle, but everybody else connected with the jet engine in its early days, thought that its civilian use would be confined to mail-carrying aeroplanes; not alone Fleming, but everybody else at the beginnings of Penicillin, failed to see its therapeutic possibilities. "Columbus, thinking he had found a new route to the East (with all that that implied) when in fact he had discovered America (with all that *that* implied) is the very type of all inventors. What the inventor does not know, the innovator has to find out. Innovation is, inescapably, a learning process."(14)

"WHOLE INFORMATION CONTENTS"

This being so, it flies in the face of reality to require the innovator to define what is new in what he has done, at the very beginning, rather than at the end of this process, and to limit his protection to whatever he is able to define then. Yet this is what we do when we grant a monopoly in terms of abstract invention Claims, rather than in terms of the "product-in-being", to borrow Kronz's phrase. It may be added that similar problems arise in other forms of intellectual property, notably in Designs, and not just in respect of Patents. In all cases, the solution can only be found in terms of a *"whole information contents"* approach.

This is perfectly possible as long as the monopoly protection is granted for innovation and not for invention, so that judgments as to information content can be made *retrospectively*. This view, it should be noted, is endorsed by some comments in his Chapter by Professor de Jong. He points out that innovation is only distinguishable *ex post*. No operational definition can be given, he claims, that would make it possible to decide *ex ante* whether an economic act falls into this category.

But as soon as an actual innovation starts interacting with its environment, a dynamic process of information production is set in motion. This process makes it possible, with the clarity of hindsight, to see precisely what it was that was new, in informational terms, in the innovation, and consequently to set the correct limits to the monopoly in the case of any dispute.

An investment decision to innovate under the classical Patent system has to be made on the basis of guesswork as to how far the Patent Claims will in fact cover any *valuable* new information which the innovation may produce. If the most valuable information does not fall within the scope of the Claims, the investment will have been made largely for the benefit of competitors. It frequently happens, for example, that by far the most valuable new information generated by an innovation, relates to the market (actual or potential) for a particular type of product. Since this is not protected by a classical Patent, the innovator who has risked his money to produce it, has put himself very much at the mercy of his rivals. If they have more Capability and/or Persuasive market power than he has, they may even reduce his market share so much that he actually ends up losing money.

In contrast to the classical Patent system, the Innovation Warrant faces up to the amount and complexity of information in modern products, processes and services. Instead of "Claims", there is first of all a description of an innovation (product or process) in respect of which investment is proposed. This description is matched against everything that is known about what is already available (including information from Third Parties with an interest in preventing a Warrant from being granted) for purposes of deciding whether or not protection is justified.

Once it is granted, however, it is not the content of this description which counts from the point of view of defining the limits of the monopoly, but *the actual innovation itself*. This, as it becomes progressively realized, takes over the function of defining its own information content. If, as may be expected, there is a substantial gain in information in the period between start of investment and first commerical product, well and good; the innovator has risked his money, and is entitled to the benefit of the consequent "learning by doing". It should be noted that although the Innovation Patent does use Claims, these are drafted retrospectively, at the stage of the "product-in-being". They can therefore take account of this learning process.

The classical Patent puts a huge premium on foresight, a premium which the statistics of failure to protect innovation show to be very rarely earned. Both Innovation Patent and Innovation Warrant are designed to encourage, not legalistic peering into the future (which is what Patent Claims drafting is) but action to *shape* the future. Both then rely upon *hindsight* to establish just what new information has been generated by this action. In the Warrant's case, the monopoly grant explicitly relates to all of this. In the event of a dispute, the innovation speaks for itself, and hindsight will show clearly whether or not its new information has been used unlawfully by another.

MATCHING INFRINGER'S ADVANTAGE

Using the Piatier analogy, for a classical Patent to be able to protect innovation well, it would require of someone close to the source, that he be able, not only to define the minerals in the water near the river's mouth, but also to specify their relative proportions accurately – an impossible task. In the case of the Innovation Patent and Warrant, because the analysis is carried on *downstream*, when the elements brought in by all the tributaries are available for analysis, the correspondence between what needs to be protected and what is actually protected can be near-perfect.

The "whole information contents" approach, in particular, simply puts whoever has generated the new information, on an equal footing with an infringer or would-be infringer in a dispute. In the classical Patent system, the advantage is all with the infringer, to the extent that hindsight (in copying another) is so much better than foresight (in trying to delimit in Patent Claim terms where the true value of an innovation will be found in the future). The doctrine of "commercial equivalence" of the Innovation Warrant, allows both parties to use hindsight equally, so that disputes are settled on a basis of the truly new information content of the innovation – and all of this. Anything less than "commercial equivalence," it may fairly be claimed, is inadequate to cope with the growth of the information economy.

This point may be granted, whilst leaving doubts in the minds of some as to whether the doctrine could be administered successfully in practice. Two points may be made in respect of this: –

Firstly, those who copy innovations seem to have little difficulty in recognising – and at lightning speed – just what is new in an innovation, if there is even a hint of favourable market response to it. This gives grounds for believing that experts (and in this context there can be no better experts than the innovators themselves and their competitors) will readily be able to recognise, firstly, what is new in an innovation, and secondly, whether any alleged infringing product is making use of new information generated by the innovation in question.

Secondly, Anglo-Commonwealth Courts have already used a very similar criterion in some Copyright cases. In these, their test of infringement was simply

"whether the defendant has unjustly benefitted from the labour, skill, effort and investment of the plaintiff – in other words, infringement by taking unacceptable short-cuts". (15)

The fact that the "whole information contents" approach can apparently be used successfully in this way for Copyrights, augurs well for its application to direct protection of innovation.

PATENTS AND PRIZES

In their Chapter, Mandeville and Macdonald laid great stress on the way in which Silicon Valley, as the paradigm of the information economy, is characterized by, amongst other things, "an important role for both small firms and individuals." In the innovation context, "individuals" are private inventors, and attention has been drawn above to the way in which the Innovation Patent, in particular, offers special advantages to these.

This point has received confirmation from a most interesting and relevant aspect of the new economics of Patents, as developed by Professor Wright elsewhere, but only referred to briefly by him in his Chapter. Part of this research involved plotting probability of success of projects against elasticity of supply of Research and Development. It emerged that there is an area where prizes appear to be the best means of encouraging the desired results. *Ex ante* prizes, where the Government specifies both an object and a reward, have been responsible for the chronometer and canned food, amongst other inventions. The best known examples of *ex post* prizes are the British Commissions on Awards to Inventors after each of the World Wars. The area where prizes appear to be most effective is where probability of success and elasticity of supply are both low – emphatically the territory of the private inventor.

The scale of the odds against commercial success by private inventors to-day is too well known to require elaboration. The "elasticity" aspect is newer. What this measures is the extent to which R&D inputs are explicitly related to the immediacy and certainty of cash returns. At the "elastic" end of the scale, for example, would be the Defence Contractor who will only perform R&D at all if he is fully compensated for it through a Government research contract. At the other end is the small residual band of private inventors referred to by Professor Piatier, whose efforts are stimulated by the prospect of immediate monetary reward only to a small degree. Historically, indeed, many private inventors seem to have been driven by forces that made them altogether indifferent to money, according to any commercial time scale. This, of course, is precisely what inelastic supply of R&D inputs means.

The new economics of Patents consequently endorse the claim that the Innovation Patent is particularly apt for rewarding inventors of this type. It will be recalled that if its novelty criterion is met, it is a moral certainty that such a Patent will be granted, since the protection it gives is limited to the actual embodiment that is commercialized. The Innovation Patent, therefore, can be envisaged as an *ex post* prize system depending only on meeting the requirement of novelty, and with the prize proportional to the innovative capacity of the applicant.

Its provisions in respect of the variable term of grant also have special value for private inventors. Since these correspond inversely to the "innovative capacity" of an applicant, a private inventor, who has little of this capacity, will get a longer term for a given innovation than any firm will. However, if a firm becomes involved as "substitute innovator", the term granted becomes the arithmetical mean of what either would receive on its own. From the firm's point of view, the fact that the private inventor can give it, not just his ideas, but also the right to a longer period of monopoly in exploiting them, is as if he were in a position to make an equity investment in the project. The result can only be that private inventors should find it much easier to get support for their ideas if an Innovation Patent were available to them, than they do to-day.

Too much stress should not be laid on this endorsement by the new Patent economics. However, it is yet another indication of how this version of direct protection of innovation can apply the principle of offering monopoly in exchange for action deemed to be of social value, in ways that go well beyond the range of the classical patent system.

REDUCING UNCERTAINTY

How would "commercial equivalence" and "whole information contents" interact in practice?

Consider first the position in which the management of a firm which contemplates investment in innovation finds itself. On the one hand, it wants to survive and to make money, and successful new products are essential to both objectives. At the same time, it faces a number of different types of uncertainty. It may not succeed in the technical aspects of any innovation it attempts. Even if it does, the market may not be ready to accept the new product, or there may be failure to devise a suitable marketing strategy. And the firm which gets all these aspects right, may still end up losing money instead of making it, because competitors are able to imitate its product successfully and rapidly. Every reason why it may be unable to obtain rewards from its investment in generating new information is a valid deterrent to making such an investment.

Of all the different types of uncertainty with which such a firm has to contend, the only one that the State can do anything about, is the last, which relates to appropriability. Even if this were to be removed completely, there would still be quite enough uncertainty in respect of other aspects of the project, for the rational decision in many cases to be *not* to invest.

But if it is public policy to encourage innovation, and if removal of uncertainty about appropriability is adopted as an instrument of such a policy, then the objective must be to remove it as completely as is consistent with the cost of whatever measures need to be taken. There does not appear to be any rational basis upon which the removal of "less" rather than "more" uncertainty (always bearing in mind the related social costs) could be justified. This point is made because tolerance of the classical Patent system in the face of all the evidence that it works badly, seems to be based upon some vague idea that it would be "wrong" to try to make it work well.

MANAGEMENT IDEAL

The next question to be considered is what would perfect appropriability mean in practice, to the management of a firm contemplating investment in innovation?

There can be little doubt that this would be absolute insulation of sales of its new product from all actions of whatever kind by competitors, without limit of time. But any public policy to encourage innovation by providing the conditions for appropriability, must temper the extent to which the arrangements meet this management requirement in any individual case, so as to maximise the number of decisions that will be made to invest in innovation.

The most obvious element of such a balance is restriction of the Patent term, since giving one innovator all he wants in this respect would prevent other potential innovators from ever acting in the same field, and so would reduce the total amount of innovation. This is quite apart from the fact that in some cases, the reward to the protected innovator would reach levels that would be socially unacceptable and which would therefore cause public reaction against the protection system.

Professor Wright has shown above how, as the economic theory of Patents has become more sophisticated, it has moved from endorsement of a static Patent term, to laying stress on the importance of matching the term of protection more precisely to what it is intended to do. This idea, of course, is fundamental to both proposals for direct protection of innovation, in which it is developed into specific arrangements for variable terms, which will be reverted to below.

MARKET INSIGHTS

If the monopoly is to be really worth having, it must cover all the new information generated by the innovation, whether embodied in the product that is actually manufactured and sold or not, whether explicit or implicit. In this context, it must be repeated that very often the most valuable new information which an innovation brings with it, relates to *the market*. This is shown by the number of cases where rejection of a new idea by established firms has been based, not upon scepticism about its technical feasibility, but because "there would be no market for it". This is the point of greatest uncertainty in the whole innovatory process, and it cannot be resolved until every one of the earlier stages (listed by Piatier) has been traversed, and the high-risk investment is approaching its maximum amount.

Classical patents, even when administered according to the doctrine of legal equivalents, totally fail to protect this most important type of new information, and their power to confer appropriability is correspondingly diminished. Information as to the market may be difficult or even impossible to articulate, and yet may have a profound effect in terms of uncertainty-reduction.

Market signals can be received in the subconscious of experienced business men, and can cumulatively build up towards making a particular decision possible. Such information, even if it is only grasped with difficulty, must be considered as part of the explicit contribution of the innovation in informational terms. It would be wholly absurd to think that information of this type could be captured in traditional Patent "Claim" terms, and yet its reality and importance cannot be gainsaid.

The object is to give a monopoly that genuinely removes the element of risk associated with "appropriability" from the total risk involved in investment in innovation. Consequently, any action which diminishes the power of the Warrant-holder to capture the reward of his risky investment fully, must count as infringement. Because the Innovation Warrant extends the doctrine of equivalence beyond "technical" or even "legal" concepts to the reality of actual business, infringement comes to mean:—

"Causing, or attempting to cause the revenue from a product protected by an Innovation Warrant, or produced by a process similarly protected, to be reduced, other than by innovation."

Faced with this doctrine, a would-be competitor might be expected to apply for his own Innovation Warrant in the first instance, since holding one Warrant, being evidence of genuine innovation, which makes use of none of the new information in any protected innovation, must always be a good defence against an infringement charge. If he fails in this, he may be tempted to put his product on the market anyway, and risk that the original innovator may obtain evidence that he has lost (or could lose) sales as a result, and so charge him with infringement.

However, arbitration (which, thanks to Professor Tullock's work, is now seen as having a vitally important relevance to any system of direct innovation protection) would settle this matter quickly and decisively. The specific provisions for arbitration that are now proposed, will be discussed below.

Only experience with the arbitration procedure will tell whether or not firms would accept that failure to obtain a Warrant for their own new product is a signal either that they should make another attempt at innovation, or hold off introduction of a product competitive with the originator's, until the latter's monopoly term has expired. Shorter terms of protection as the general rule, will help here. The "competitor" firm also has an interest that the spirit as well as the letter of any regime of innovation protection should be respected, since this firm, too, may one day be an innovator in its turn, and will then want to be able to appropriate the rewards of its own risky investment without being harassed. "Good innovators make good imitators", as David Landes expressed the historical lesson. What is now needed is a legal and economic environment that will direct creative energy into innovation rather than imitation.

In a case of Warrant infringement, what would therefore have to be decided would be: Does the alleged infringing product depend to any extent at all upon information which is available solely as a result of the existence of the protected innovation? The latter would have to be given particular credit for any reduction of uncertainty as to the potential market, and since precise information on this will not be available, the innovation should be given the benefit of any doubt.

REACTIONS FROM ESTABLISHED FIRMS

Just as innovators would like the term of their monopoly to be unlimited, they would also like the definition of "commercial equivalence" to be absolute. However, some restrictions come to mind immediately as being necessary to protect the legitimate interests of Third Parties.

Firstly, what about the reaction of those firms already in the market, whose products may be expected to be adversely affected by the innovation? If they are blocked by the latter's monopoly protection from correspondingly upgrading their own products technically, they are likely to react by intensifying their selling activity, cutting price, improving their service, spending more on advertising, etc.

In trying to keep up sales of the older products, this activity is certainly "attempting to cause reduction in revenue" to the new one, but it would surely be wrong to categorise it as infringement. At the same time, it can be envisaged that under certain conditions a large, multi-product firm might decide to protect a threatened product by massive cross-subsidies from other profitable products and/or from reserves. If the innovating firm was weak (and even an initially strong firm could easily have been weakened by higher than anticipated costs of the innovation) the result could be a forced sale of the innovation, or even a take-over of its sponsoring firm.

Predatory commercial activity of this type would certainly discourage investment in innovation, and it clearly illustrates the lèse-majesté referred to in Chapter IV by setting out to defeat an express objective of public economic policy. It would generally involve a clear and identifiable attempt to dominate the Specific market power of an Innovation Patent or Warrant, by either or both of the other two market power types, Capability and Persuasive.

Since one of the main reasons for establishing a regime of direct innovation protection is specifically to put an end to the situation in which virtually all innovation has to be underwritten by these types of market power, any such attempt would have to be resisted strongly. This would especially be the case in the early days of the new system, before its merits had had time to be experienced fully in practice by the business community.

Any new system would therefore need to be armed with means of protecting innovating firms from predation of this type, whilst at the same time permitting normal marketing responses on the part of products which were on the market before the innovation. Competition Policy constantly deals with this type of problem, and it should be possible to take over realistic and administratively feasible guidelines from it.

IMPLICIT INFORMATION

Secondly, in addition to the information an innovation generates *explicitly*, in the sense that those concerned advert to it at the time, it also generates *implicit* information. This will only emerge to view over time, probably as a result of other information being brought to bear on the innovation. Examples are a process which is developed for one purpose, and which is found to be applicable also to a quite different one, or a new material whose special properties are of value in some new field which may be far from that of its first intended use. Should the innovator's protection extend to the new use also, having regard to the fact that he never dreamed of it? The Innovation Patent answers this question negatively, limiting its "Option" claims to variants of the "Copy" claim embodiment that have been tried and tested by the applicant.

The Innovation Warrant, in contrast, would grant the extended protection, for three reasons: Firstly, the administrative simplicity of a definition which refers to *all* new information, would be of great value. Secondly, it is in the very nature of innovation that its full implications are only drawn out over a considerable period of time; it does not seem to be in tune with this reality to pay attention only to those aspects which can be grasped in advance, or at least quickly. Thirdly, holding out the possibility of an uncovenanted reward (or rewards) to an investor, may compensate to some extent for all those elements of uncertainty that cannot be removed from his decision to invest, and so make such a decision possible more frequently than would otherwise happen.

In principle, therefore, the Innovation Warrant would protect the new information of an innovation against being used without permission, even in a context which the innovator had never foreseen. But the overriding provision remains that infringement must affect, or attempt to affect, the innovator's revenue. If he is not actively exploiting the information in the new area, he will have no revenue which could be adversely affected, and so others are free to use his new information in this market.

However, since they would become open to a charge of infringement once the innovator himself began to exploit it, they would be unlikely to enter this market, except under licence. It is evident that the pressure of the system is all in the direction of having the information exploited in the new way, either by the original innovator or by others.

The innovator can appropriate the full benefit of all possible uses of the new information he has generated, but only if he is willing to make the appropriate investments for its exploitation. Otherwise, the information is open to all. There would be little administrative difficulty in ensuring that an innovator did not sterilise a particular new area of application of the information, by merely pretending that he proposed to invest in it.

INCREMENTAL CHANGES

Thirdly, what about an incremental change to a protected product, brought about by some agent outside the original innovating firm? The difficulty that arises here is of reconciling the principle that all new information arising from the innovation is to be protected, with the reality, so trenchantly endorsed by Mandeville and Macdonald, that "a single firm is unlikely to be able to improve a technology to the same extent as a vigorous cluster of competing firms".

One possible solution would borrow from British pre-1977 Patent Act practice, and would be all the more feasible because terms of direct protection would tend to be significantly shorter than for classical Patents. According to this, protection would be granted to the second innovator for the incremental change, but the grant would be endorsed with the caveat that the innovation could not be put into practical effect without infringing the first innovator's rights (since it would use some of the information generated by the first innovation).

The position would then be that use in practice of the incremental improvement would depend upon agreement between the parties. The first innovator will naturally want it to be incorporated in his product as soon as possible. The second innovator has a choice between coming to an arrangement with the first innovator to permit this, so benefitting from his innovation immediately; or of waiting until the first innovator's protection has expired, when he would be free to make and sell the primary innovation with his own improvement incorporated in it (only the latter aspect would then be protected).

It is reasonable to assume that prudence would force a decision in most cases in favour of reaching some mutual accommodation, so that the rapid rate of technological change that Mandeville and Macdonald rightly see as the result of a multiplicity of decision points, would be achieved.

RADICAL INNOVATIONS

A contingency against which an investor in innovation would like to be guarded, concerns further radical innovation which owes nothing at all to the first one. The frequency with which such innovations appear in any particular field tends to be measured in decades rather than single years, so that the danger that the sales of an earlier innovation could be affected adversely by a second one during the latter's period of monopoly protection, is slight. Yet it cannot be denied that it does add another element of uncertainty to the investment decision. Even though the probability of the occurrence may be low, its consequences could be catastrophic.

Although its existence ought indeed to be recognised, this is one of the elements of uncertainty in all investment in innovation which the State should not attempt to remove. Refusing to do so would be more of a symbolic gesture than anything else, because in practice a conflict of this type would be extremely rare. Nevertheless, such a gesture would show that public policy favoured innovation generally, even at very occasional cost to individual firms or investors.

This in turn, would call the attention of managements to an important point which is missing from the general innovation literature, but which was raised above by Piatier. He quotes Tessier du Cros with approval to the effect that the life of a firm is "a succession of projects" (or innovations). From this point of view, it can be considered to be more valuable to any individual firm, never to be in danger of being blocked from making an investment in radical innovation, than that it should be correspondingly secure from the very slight danger of having to meet similar competition itself, if it does so successfully.

Further, since by definition the second, radical innovation is neither available in the ordinary course of trade nor uses any protected information, it must be entitled to protection itself, and thus would consequently be free of the endorsement on its grant, referred to above in relation to incremental innovation. The legislation should explicitly provide that this freedom would be an automatic defence against any charge of infringement by the earlier innovator.

SUMMARY OF CONDITIONS

It may now be helpful to summarise the prospects for protection of a particular innovation, as these would look to a potential investor in it:–
1. Exclusive commercial use of all new information arising from an actual innovation will be granted for a prescribed term.
2. "Commercial use" is defined as "use which affects, or is intended, or could reasonably be expected, to affect revenue from the innovation".
3. Once granted after appropriate provision for objections, the only ground on which this monopoly can be challenged is that it had been obtained fraudulently.
4. Exclusivity of commercial use of the new information would cover applications of it which were not foreseen by the innovator at the time of the grant.
5. Incremental innovations can be the subject of protection granted to others

to the extent that they generate "new information". But, whenever in operation they also use information that is the subject of an earlier grant of protection, they cannot be put into effect before the expiry of the latter grant, without a licence. This would be endorsed on the grant of protection for an incremental innovation.

6. Radical innovations, which use no new information that is protected by a grant of exclusivity, may be freely commercialised, even if doing so has an effect on the revenue of an earlier innovation.

7. A grant of protection which carries no endorsement to the effect that its subject-matter is an incremental innovation as defined in point (5) is a valid defence against a charge of infringement.

It is claimed that such a system, properly enforced, resolves the question of appropriability as far as it is reasonable to do so. A prospective investor does not have to face imitation of, or "inventing around" his innovation; nor will others be able to exploit a hitherto unknown market which he has uncovered, if he himself is doing so. If his courage in venturing into the unknown results in discovery of a gold mine which he never dreamed was there, he can still get the benefit of it by extending his commercial activity appropriately. He is placed in a reasonable negotiating position in respect of incremental improvements to his innovation by others.

On the other hand, no firm or individual is prevented from using every scrap of new information in ways that could lead to still further innovation. Its "commercial use" is indeed barred to them for a time, but then, they had done nothing to bring that new information into existence. They can make use of it in any way that does not affect the innovator's revenue, for example in R & D for an even better product, to be put on the market the moment his protection term expires. If the innovation can be put to use in some way which had not been foreseen when it was launched, they may exploit it in any new way to the extent that the innovator fails to do so himself.

Finally, they are of course always free to respond to an innovation by themselves trying to produce a further radical innovation. This would owe nothing to the information generated by the earlier one, and could therefore be exploited without any reference whatever to any damage it might do to the first innovation's revenue. Solid reassurance on this freedom to exploit could be obtained simply by applying for, and being granted protection in turn.

THE TERM OF PROTECTION

It is a common feature of both proposals that the monopoly period should not be fixed, as it is in the classical Patent system, but should be matched to the task it has to do. The Innovation Patent seeks for perfection in this, and treats each individual case on its merits. The Innovation Warrant uses categories which are based upon risk, both as measured objectively (in so far as this is possible) and as seen subjectively by investors.

The variable term receives support from the new economics of Patents which are outlined by Professor Wright in his Chapter. However, these naturally assume that the system they are examining is effective in achieving its stated ends; consequently they depend upon the premise that

"the entire marginal (but not total) benefit of the discovery is appropriated by the holder of the patent during the term of the patent".

As has been discussed earlier, particularly in relation to the Mansfield and Yale studies, this is manifestly not the case with classical Patents. Consequently, the "Common Pool" problem, concern with which distinguishes the new economics of Patents from the old, scarcely arises in practice. It is only if too much R & D were to be done, because of the reward of monopoly that is held out as an incentive, that this becomes an issue.

Classical Patents offer such a poor incentive for rational investment in R and D, that this is one charge at least, that cannot fairly be levelled at them. Further, the "Common Pool" problem reflects the fact that only one of a number of firms which had invested resources in R & D could obtain the (real) monopoly. Since this monopoly would "block" the others, their efforts would have been wasted. As empirical observations of the workings of classical Patents show, this is anything but the case in practice. The opportunities for "inventing around", and even for direct imitation, are so great under the present system, that any investment at all in R & D that results in something the market wants, is highly unlikely to be prevented from obtaining a reward by the existence of a classical Patent grant to another. Nobody minds a Patent term being 17 or 20 years, since it is not an effective monopoly anyway.

"STRENGTH" VERSUS "LENGTH"

However, the situation would be quite different with direct protection of innovation under either of the systems proposed, since these would really be delivering "appropriability". In both, the monopoly grant, once made, can only be questioned on the ground that it had been obtained fraudulently. This immediately removes the usual ground of counter-attack in infringement actions, which is to claim that the Patent is invalid. It correspondingly strengthens the position of the holder of the monopoly. The arrangements proposed for Arbitration and Litigation (see below) do the same, especially for small- and medium-sized firms, because there would be inexpensive and speedy decisions. The "total information contents" approach to infringement of the Innovation Warrant (although not of the Innovation Patent) would complete a "package" which would bring the actual state of affairs much closer to the theoretical one dealt with by Wright.

Consequently, his conclusion that a variable term must replace a fixed one, supports the proposals. Attention must then also be paid to the indications from his theory that –

1. Optimal monopoly life is much shorter than has been the case with actual Patent systems, and

2. This optimal life is "very sensitive to parameters that are difficult to identify".

(1) is in line with the assumption in both proposals that the strengthening of monopoly grants which direct protection of innovation offers, would be associated with shorter terms than those of classical Patents. It is believed that innovators would welcome these if they meant real protection, and that public opinion would also find shorter terms an acceptable counterbalance to stronger monopolies. (2) lends support to the Innovation Patent's ideal of *individual* matching of the term of grant to "innovative capacity" on theoretical grounds, even if this might be difficult to achieve in practice.

MEASURING NECESSARY TERM

It was the anticipated practical difficulty of administering a variable term that was the source of most of the criticism of this aspect of the proposals. Silberston, in particular, considered that both of these "greatly underestimate the difficulty of arriving at an appropriate length of term". The Innovation Warrant's way of dealing with the problems of measurement is by combining project-related and firm-related risks. de Jong holds that it is no business of the State's to differentiate between its subjects, and so he rejects the idea of taking risk of the "firm-related" type into account. Wright, too, objects to this, on the ground that it would tend to prevent innovation being undertaken in whatever size of firm is most appropriate for it.

In defence, it may be said that the "firm size" element in the risk matrix suggested for the Warrant is no more than a surrogate for risk as viewed *subjectively*. It is agreed that if we had objective measurements of risk, there would be no need for it, but the problem about innovation is precisely that such an objective measurement is lacking. The more radical the innovation, the more it involves uncertainty rather than risk, and the more this lack will be felt. The intention is to encourage investment by removing as much as possible of one element of the uncertainty which an investor faces (that relating to the appropriability of the results, if there are any).

It is therefore uncertainty *as it it experienced by the potential investor* that is important, not as how it may appear to someone whose money is not at stake. Uncertainty may even look like risk to someone who does not stand to lose!

As the Innovation Patent explicitly provides, the ideal would be some form of matching in each individual case. Such matching would be independent of firm size, since one (large) firm's subjective assessment of the risk in any particular case may rate it higher than another (small) firm would. Because of this independence, the ideal arrangements might be acceptable to de Jong and Wright, but they are hardly practicable with our present resources for measurement.

The fact that we cannot measure well enough to-day, however, does not mean that we will never be able to do it. Computerisation of every aspect of the work of more and more firms seems certain to transform our capacities in this respect. All that is necessary at the outset, for both Innovation Patent and Innovation Warrant, is that the *principle* of a variable term, based upon empirical research into the risks, both objective and subjective, of investment in innovation should be accepted.

de Jong favours three terms only, presumably relating to incremental innovation, path-breaking innovation, and an intermediate type. This would be a perfectly sound working basis on which to make a start, but it would also be essential to impose a duty on the Innovation Office from the outset, to improve the sophistication of the matching of its terms to risk. Corresponding to this duty, it should be given the resources to carry out substantial and continuous empirical research to this end. Eventual arrival at the ideal of variable terms that were individually appropriate to each case, and still acceptable generally, need not be excluded.

PUBLIC OPINION

Silberston questions whether monopolies of the strength proposed would meet with public approval, citing the present objections to certain pharmaceutical Patents. It would obviously be easier to gain this approval, with monopoly terms that were significantly shorter than the public now takes for granted in the case of classical Patents. The Innovation Office's researches and the resulting better matching of monopoly term to the need for it, could only improve the level of public acceptance of direct innovation protection.

In the broad sense, of course, public opinion is not concerned in the slightest with innovation protection. The general public knows nothing at all about Patents, and very little about monopolies. Consequently, when the word "public" is used in this context, what is meant is "the business lobby". This, in turn, will be activated, not by innovators, but by their less dynamic competitors, who would naturally prefer to be free to invent around, copy and plagiarise the work and efforts of others, whilst using legal devices to postpone Court decisions until these have become largely irrelevant.

Pressure from this quarter must be resisted. It is short-sighted, in that the mass of firms which follow rather than lead, would find that there would be many new areas opened up for them to operate in, if direct protection of innovation existed. A relatively small number of pioneers would then enjoy a period of insulation from competition, and so could break new ground, which would eventually be available for others to exploit also.

Genuine innovators in the business community would welcome the change from long terms which are largely worthless, to short, valuable ones. Even if this means that they may find themselves blocked from participating in a particular innovation, in which they would have liked to be involved, they know that when it comes to their own turn, they will be glad that the system does deliver "appropriability".

Professor Tullock referred in his Chapter to the "stochastic process" aspect of innovation litigation. His valuable insight can in fact be extended with great advantage to the whole innovatory process. What matters most for any system of innovation protection is that it underwrites a stream of innovations over time, so that for any individual firm, its aggregate gains will far outweigh any occasional loss.

CASE STUDIES

The issue that Silberston has raised, however, is an important one, and must be examined from every possible aspect. What are only theoretical issues for the classical Patent system, because it is so ineffective, could become genuine problems for direct innovation protection. Because the classical system does not underwrite enough R&D, the question of the "Common Pool" does not have to be considered in its practical economics; because the barrier to entry to a market which it raises is so low in practice, there is no incentive to protest against its time-scale in theory.

But, just as it has been noted above that the theory of the new economics of Patents is relevant to direct protection of innovation, because this would provide a strong incentive for R and D, so the length of the monopoly would become important to all sorts of firms which can largely ignore this aspect of classical Patents. This would be all the more so, because the grant would be incontestable, except in the rarest cases, and the Arbitration system (to be explained below) would make enforcement easy, certain and quick. How likely is it that objections to the strength of direct protection could multiply, as Silberston fears?

At present, we only have one way of getting an insight into this problem, which is by inference from classical Patent case-histories where the link between invention and innovation is close. One of these is Xerox (United States Patent No. 2,297,691, issued on October 6, 1942 to Chester Carlson). The inventor, who was himself a qualified Patent Agent, claimed for the use of Selenium in "electrophotography", and nothing was ever found to replace this during the life of the Patent.

In this case, therefore, indirect protection of innovation (direct protection of invention) operated just as direct protection of innovation would have done. Because this was almost unique for an electro-mechanical invention, for the purposes of the present study, Xerox were asked to provide historical information on their sales and profits, and they generously complied.

It was because of the existence of his Patent that Carlson was able to interest the Battelle Institute in investing to develop his invention, in 1944. Because of it, too, in turn, the small photographic firm, Haloid, staked its future on "electrophotography" in 1946. It took three years to produce the first product for the market, and a further ten before the real break-through commercial product, the Xerox 914, was launched, in 1960. By some measurements, this machine ranks as the world's most successful commercial product.

VALUE OF XEROX PATENTS

The first indicator of the value of the basic Xerox Patent and a few other Patents of lesser importance issued to Carlson, can be found in the growth in value of Battelle's interest. These basic Patents had been made over to the Institute by Carlson subject to a royalty, on the condition that Battelle invested $3,000 in developing them. Twelve years later, when they still had about three and a half years to run, Haloid bought them for shares estimated at that time to be worth $3,800,000. It cannot be assumed, however, that this represented an appreciation in value of 1000%, because there were certain tax advantages for Haloid in this purchase. In spite of incomplete figures, the pattern of evolution of this firm's sales and profits is clear, and shows for how long rewards were in fact postponed: –

HALOID CORPORATION ($million)

YEAR	SALES	CASH FLOW	PROFITS
1947	7		0.125
1948	8.6		0.449
1951		0.719	
1955		2.045	
1956	26		
1957			
1958	27.5		
1959	33		2
1960	37		
1961	59.5		
1962	104		
1963			
1964	268		
1965	392		58

Carlson's original Patent expired on 6th October 1959. By far the most striking feature of the figures is that sales and profits only started to grow very rapidly *after* that time. In fact, it is clear that in this case, which is certainly one of the classical Patent system's prize exhibits, the result achieved *within the Patent term* was less than a five-fold increase in sales. Considering the risk involved at the outset, a profit of $2 million in the year when the Patent expired, could hardly reflect any kind of excessive reward from holding a Patent monopoly. And if this is the case with Xerox, how much less are profits likely to be for run-of-the-mill innovations?

It is, of course true that the protection obtained from the Patent system by Xerox was by no means limited to the basic Patent. By 1950, they had 30 secondary Patents, by 1958, 126. 33 Patents were obtained in 1961 alone, bringing the Patent "position" which had been built up by then, to around 200. This "position" is undoubtedly associated with vastly higher levels of profit than the original Patent ever was.

But two other factors had by then become at least as important. These are the capability which the firm had built up through many years' experience and investment in R and D, on the one hand, and its marketing strength on the other (the decisions to sell copies rather than machines, by leasing the 914, for example, and to develop outstanding showroom facilities, were critical). The huge profits earned during the 1960s and later, therefore, are probably more attributable to Capability and Persuasive market power, even than to the Specific market power of the Patent "position".

POLAROID CONFIRMATION

This conclusion is substantially supported by the history of Polaroid, a firm which was built on the explicit policy of never making or selling any products unless they were covered by Patents. Edwin Land's first Patent application was filed in 1928, and by 1935 he held ten Patents on light-polarizing materials. The first application in respect of the Polaroid camera was filed in 1945, and by 1960 the firm held about 300 Patents in this field in the U.S. alone, and another 400 in 24 other countries. Like the Xerox Patent, these Patents of Land's were of the kind that gives real protection, as is evidenced by the fact that firms such as Eastman Kodak and American Optical had to take licences under them from 1934 onwards. The Patents also largely underwrote the external financing by bankers of Polaroid Corporation. In the period 1927–34, Land operated as a private inventor, with financial support from his family and friends. The first commercial sales of his synthetic light polarizer material were made in 1934, and further development was financed from these until the formation of Polaroid Corporation in 1948. During this period annual sales were between $3 million and $5 million. In 1949, they were $6.5 million, but ten years later they were $90 million. (16)

The Polaroid camera is to light-polarizing, what the 914 Copier was to "electrophotography". As in the Xerox case, the main revenue flow did not begin until after expiry of the basic Patent. And it seems equally certain that this flow was largely due, not to any single Patent, but to a Patent "position", which cannot be separated from both capability and persuasive market power.

In both cases, other firms which would have very much liked to be involved in the developments, were effectively excluded, and the resulting profits were very high indeed. Yet there is not the slightest evidence of calls for abolition of the Patent system as a result. This suggests that the objections to pharmaceutical Patents to which Silberston refers, relate to the *subject-matter* of the protection, not to its strength. If this is indeed the case, the strength of the grants which would protect innovation directly in other fields, is correspondingly unlikely to be a source of contention.

Although more research would be needed to confirm the point, there seems to be a second, even clearer lesson from both the Xerox and Polaroid cases. This is that if there are objections on political or social grounds to monopolies granted under the Principle of Patenting, there is no point in expressing these as a demand for limiting the term or strength of any *individual* grant. If the master Patents in both these important cases resulted only in relatively small profits during their 17-year terms, no single non-chemical classical Patent *on its own*, can ever have produced profits that could be regarded as excessive.

Very large profits in both cases only appear with the grouping of a number of Patents into a "position", and this is also invariably associated with the build-up of capability and persuasive market power. In the case of Drug Patents, the individual grant may be more significant relative to a Patent "position", but here, too, the other types of market power are of the greatest importance. Marketing techniques, as pointed out earlier, are particularly effective for pharmaceutical products, because they are directed at doctors, who do not have to pay for what they prescribe.

Two conclusions therefore emerge with substantial certainty:–

1. It would be quite perverse to attempt to curb the value of grants made under the Principle of Patenting, whilst leaving capability and persuasive market power untouched.

2. If, as part of a general reduction of market power on policy grounds, it was considered desirable to reduce the value of classical Patents, Innovation Patents, or Innovation Warrants, nothing could be easier. All that would be necessary would be to limit the number of such grants which could be held by a firm simultaneously, in respect of any single area of technology. The target would therefore be the "position", comprised of interlocking monopoly grants, rather than any individual grant.

RESIDUAL POWER

As a matter of practical politics, the authorities would no doubt favour arrangements which reserved to them some ultimate power of bringing a grant to an end, even if this was intended to be used extremely rarely, if at all. Indeed, the only imaginable occasion when this power might be exercised would be when a firm had stumbled upon a development with so many possibilities that the resources of all the firms in an industry, and not just a single one, ought to be brought to bear on them immediately. An example (which in itself shows how hypothetical the issue is) might be if a small firm, and not Bell Laboratories, had invented the transistor.

How could such a residual power be reconciled with the incontestability that is a feature of both Innovation Patent and Warrant? It is vital to preserve this, since even the slightest possibility that a grant could ever be revoked would have an altogether disproportionate effect upon investment in innovation. One way out of the dilemma would be to use a fiscal device. In the rare case where the Innovation Office became convinced that the public interest would be best served by prematurely ending a monopoly it had granted, it could recommend to the Government that it make an offer of a period of tax relief on profits from the innovation, in exchange for extinguishing the Innovation Patents or Warrants.

This preserves the incontestable character of the grant, because there would be no question of compulsory licensing or purchase. The Government would, in effect, be offering to buy out the monopoly, but not for a specific sum. The investor in the innovation would be offered a new possibility of a stream of tax-free profits extending into the future, whose size would partly reflect a reward for the success of his earlier investment in innovation and partly his skill in developing the innovation further, but in the context of competition, not monopoly. He would have to decide whether he would make more money from taxed monopoly profits, or from facing competition, but with no taxes to pay.

FISCAL DEVICE

The crucial question, of course, is what should be the length of the period of tax freedom? If this can be worked out properly, the originating firm should be able to foresee that it would make more with tax relief and competition than from keeping its monopoly and paying tax; and the Government should make at least as much in additional tax revenue from the firms that would be admitted into the otherwise protected area, as it would forego in respect of the profits of the originating firm. Evidently, no such "buy out" could take place until an innovation starts to contribute to taxable profits, which means until all the investment made in it had been earned back. (This also presumably means that much of the protection period has been used up).

To avoid complex accountancy, it would be desirable to operate on some assumption which would allow sales of the innovatory product as some proportion of the total sales of a firm, to stand for the corresponding proportion of profits. It would not be right simply to assume that profits were in the same proportion as sales, since the innovation could be taken to be more profitable at the stage in question than established products. Profits as calculated for tax relief might therefore be taken to have twice or even three times the share of actual profits, that sales of the innovation have of total sales. In other words, if sales of the innovation represent 20% of total sales of a company, tax relief might apply not to 20% of profits, but to 30% or even 50%.

Fiscal means are very widely used as incentives for action directed to different economic objectives. There is no reason why they cannot be combined effectively with direct protection of innovation so as to achieve a final element of flexibility in the monopoly term, however rarely there might be need to use it.

LITIGATION

Although conceived independently, the Innovation Patent and the Innovation Warrant, in their first formulations, had a number of common features, and the present work has been able to extend their range of agreement. One of the few remaining areas of difference relates to infringement. The monopoly grant of the Innovation Patent is defined in terms of "copy" and "option" claims. Infringement of a "copy" claim would be a very unusual event, since such a claim reads on the actual embodiment put on the market by the Patentee, in all its concrete details, but with protection extending to their technical equivalents only.

"Innovating around" such a claim, by changing the product in ways that are not technically equivalent, even though they may be so in a legal or commercial sense, is indeed a course which could be followed by a typical competitor. But if this course did appear to be difficult or dangerous, he could choose another. It is open to him to seek a licence under one of the "option" claims. These read on alternatives which have been considered and tested by the innovator, and a licence to use such alternatives *must* be granted. Consequently, these claims, too, are extremely unlikely to cause litigation.

Presumably there could be rare cases where a dispute might arise as to whether an option claim applied or not, but any such disagreement would be virtually certain to be settled by issuing a licence. Infringement, therefore, may be regarded as a generally unimportant feature of the Innovation Patent regime. In contrast, it is very important for the Innovation Warrant, especially because this uses the "Total Information Contents" approach in defining its monopoly grant.

RESPONSIBILITY FOR ENFORCEMENT

It is also only the Innovation Warrant proposals which provide for lifting the burden of enforcing the monopoly grant from the shoulders of its beneficiary. The Innovation Patent takes it for granted that enforcement will remain the Patentee's own responsibility, and that such few infringement actions as there may be, will be pursued through the Courts in just the same way as in the case of classical Patents.

The suggestion that the State should actively enforce the monopoly grants it makes, did not find favour with Professor Tullock. However, his criticism, together with some very original ideas from his book, *Trials on Trial: The Pure Theory of Legal Procedure*, has in fact shown what could be the optimum way to achieve the Innovation Warrant's objectives in this respect.

The problem of litigation in any form of industrial property, is essentially one of the balance of economic resources available for it. If the holder of a monopoly can only enforce it by going to Court, and if he lacks the resources to do this, then the "protection" he is supposed to have been granted to enable him to achieve some public purpose, is effectively worthless. He is at the mercy of every economic predator. All the effort and care taken by the granting Authority to ensure that he has met the various requirements for a grant (e.g. utility, inventive step, etc. in the case of a classical Patent) is utterly wasted. An infringer, especially one who is known to have no shortage of resources for litigation, can act without fear of being challenged. At the other end of the scale, as Professor Wright has pointed out, "for those holding large financial resources, even patents of dubious validity may be very strongly protective".

The existence of any system which grants specific monopolies, is a deliberate attempt to change whatever balance of economic power would otherwise exist. It is clearly illogical to establish a monopoly system, only to have its effects largely nullified by the very imbalance in economic forces it has been set up to change. This illogicality is evident in the operation of the classical Patent system, which, as was stressed in chapter I, has virtually become a reinforcement of the other types of market power possessed by the largest firms.

This is why the Warrant proposals contain suggestions for redressing the balance of power to litigate, which at present makes Patents virtually worthless to small firms, and of limited value even to medium-sized ones. The lèse-majesté argument, that the State owes it to itself to protect any privileges it grants, reinforces the practical need to tap the innovative capacity of such firms. This argument commended itself particularly to Professor Piatier.

UNLAWFUL USE OF INFORMATION

The question of infringement is one of whether new information generated by one party, has been used unlawfully by another. This is so whether the information in question is defined in terms of "Claims", as in classical Patents, or whether the "Whole Information Contents" approach is used, as proposed for Innovation Warrants. Now, as Tullock has observed above, this is a matter of fact and not of law.

The very high cost of Patent litigation is partly due to the way in which we rely for decisions on people – the judges – who are experts in the law but laymen so far as the subject-matter of the litigation is concerned. They consequently have to be "educated" in this subject-matter, up to the level where they are supposed to be capable of giving a sound judgment on the issue between the parties. This "education" is doubly expensive, because it has to be carried out by both parties to the case; obviously, if only one party was responsible for it, the judge's "education" in the Art could not be expected to be objective.

The "adversary" system of litigation, which in general is restricted to the Anglo-Saxon countries, is consequently relevant to Patents, even in countries which favour an "inquisitorial" system in other aspects of their legal practice. But where any issue is one of fact, Tullock argues, the "inquisitorial" system gives far better value for money than the adversary one:–

"Assume, for example, that in the average American court case, 45 per of the total resources are invested by each side and 10 per cent by the government in providing the actual decision- making apparatus. This would mean that 55 per cent of the resources used in the court are aimed at achieving the correct result. Under the inquisitorial system, assume that 90 per cent of the resources are put up by the government, which hires a competent board of judges (who then carry on an essentially independent investigation) and only 5 per cent by each of the parties. Under these circumstances, 95 per cent of the resources are contributed by people who are attempting to reach the correct conclusion, and only 5 per cent by the saboteur. Normally we would anticipate a higher degree of accuracy with the second type than with the first."(17)

In the adversary system, it appears that diminishing returns set in quite early, with the bulk of additional resources brought to bear by one side, serving only to cancel out similar resources on the other. "It seems likely," Tullock observes,

"that we are indeed improving the accuracy of the court procedure when we increase the resources put into the adversary proceeding. Unfortunately, the increase is a very slow one...I would anticipate only small improvements in accuracy from quite large increases in resource investment."(18)

ARBITRATION

But how can we get over the problem of bias in "educating" the judge, if we move to an inquisitorial system? The answer must be not to use a judge at all, but instead an expert arbitrator who has no need to be educated. This cannot be done by employing experts in different fields in the Innovation Office. Each of these could still be expected to require some special "education" to enable him to deal with a particular case. This would bring us back to an adversary system, if this "education" is not to be biased.

On the other hand, if we are prepared to recruit arbitrators on an *ad hoc* basis, it should be possible to find experts in the business and technological communities, who are completely familiar with the state of any Art, right up to the point of the new information that is the subject of dispute. Consequently, they need no education in order to understand perfectly the cases made by both the innovator and the alleged infringer. Further, these cases can be made to the expert arbitrator by the parties themselves, in their own technical language, so that specialists in translating complex ideas into language that laymen can understand (the lawyers) can be dispensed with.

Tullock points out that reliance upon a single, Court-appointed "expert witness" is normal in Continental Europe, although exceptional in the United States. This practice

"amounts to taking part of the judicial power and transferring it to someone who is not a regular judge...It would make no great difference if he were formally part of the judicial staff rather than a witness".

There are many precedents for doing just this, in different forms of arbitration.

An arbitration system could also act very rapidly, in contrast to classical Patents, where the delays involved in litigation are great. Worse, they can be used by skilful attorneys to nullify in practice any monopoly a Patent purports to confer. All the time during which infringement can be persisted in because of these delays, contributes to building up the infringer's capability and persuasive market power, so that even if he eventually loses the case, the damages he has to pay may be regarded as no more than the cost of entry to a market, and generally a rather low entry fee, at that. Arbitration would put an end to the situation where time is on the side of the infringer.

ELIMINATING IMBALANCES

Its most important advantage of all, however, could be that it goes a long way towards eliminating the imbalance due to different financial resources for prosecuting disputes. Another insight of Tullock's is relevant, also developed with seemingly irrefutable geometric and algebraic logic in *"Trials on Trial"*. Because of the "cancelling-out" element in the resources used by both parties in litigation, he points out,

"we could always, by reducing the resources being invested in one side of the case, obtain the same probability as we can obtain by increasing the resources on the other; hence it is always possible to obtain the same probability of a correct outcome by a cheap method as by an expensive one."

What this would mean in practice is that when a dispute between a poor firm and a rich one goes to arbitration, all the resources of the latter have to be left outside the expert arbitrator's door. None of the rich firm's possibilities for intimidation, for hiring the very best lawyers, for multiplying the costs and extending the delays of litigation to force a poorer opponent to give up, can be brought to bear.

Large firms will obviously be reluctant to forego use of their resources in this way, and since the arbitration system would have to be incorporated into existing legal arrangements, there must still be provision for appeal to the Courts. How then can arbitration be prevented from becoming merely a form of preliminary hearing, with the imbalance it is intended to remove, remaining as strong as ever in ensuing litigation? An answer is to be found in adapting yet another idea of Tullock's, tentatively formulated originally in connection with divorce proceedings.(19) This is that financial aid should be available from the State for litigation, *but only to a party that has co-operated fully with an arbitrator, and accepted his ruling.*

THE "TULLOCK SANCTION"

Such a provision would be an extremely powerful force making for acceptance of arbitration by all concerned. No small firm would ever appeal to the Courts, since it would lack the resources to litigate. There are five good reasons why large firms, which would have these resources, would be very reluctant to use them for this purpose:–
1. The Court, which is inexpert to a greater or lesser extent in the subject-matter of any particular case, would certainly give a lot of weight to the arbitrator's decision, since he has been chosen precisely because of his expertise. The odds must therefore be against winning the legal battle, having lost the arbitration.
2. In present circumstances, even the threat of litigation will almost certainly force a small firm to capitulate. It is quite a different matter when the small

firm is given resources to defend itself (not because it is small, note, but because it has accepted the arbitration). Many large firms would not want to fight an opponent who is now effectively "their own size.'

3. Most litigation is entered into in the expectation that it will end in a compromise. Since the firm which has accepted the arbitration does not have to pay its legal costs, there is no pressure upon it to settle out of Court. Appeal from arbitration, therefore, has an "all or nothing" aspect which increases the amount of money at stake. This factor reinforces that relating to the odds against winning, making a decision not to appeal all the more likely.

4. Large firms might also be reluctant to be seen using their strength against small ones, when an independent expert arbitrator has ruled in favour of the latter.

5. Arbitration decisions would constitute a stochastic process. In his Chapter above, Tullock pointed out that for a firm which is engaged in a constant series of innovations, the availability of quick and inexpensive decisions is of the highest value, since it can expect to win some and lose some. But what is far more important, for small firms as well as big, is the pattern over the long run, not the outcome of a single "life-or-death" case. This is in line with the whole object of both proposals for direct protection of innovation, which is to contribute to making successful innovation commonplace, instead of exceptional. Tullock's insights show how the idea of "declaratory judgments", advanced in a partially formulated way in Chapter IV, could be achieved in practice at minimum public expense.

For all these reasons, it seems reasonable to expect that under the Scheme proposed, litigation would be extremely rare. Consequently, it would not be necessary to budget for large sums of public money to be used for legal aid. This, and the cost of arbitrations, would be an alternative use for a trivial part of the funds currently devoted to the direct, "interventionist" support of innovation. This type of support, as Professor Piatier observed, has proved to be singularly ineffective.

Arbitration would be just as applicable to Innovation Patents as to Innovation Warrants, and would enhance the practicality and value of that variant of direct innovation protection every bit as much. Moreover, although arbitrations would not cost either party anything more than whatever is involved in preparing its evidence and attending at a hearing, their numbers should be easily manageable. This is because of the "investment" aspect of both proposals for direct protection of innovation. To be a party to an arbitration, a firm would either have made, or be committed to make an investment (so that it could be the holder of an Innovation Patent or Warrant, on the one hand) or, on the other hand, to be a putative infringer, or want to make an investment (in something which might infringe).

In the latter case, it would seem reasonable that a potential investor who wins an arbitration, but does not then go ahead with his investment, should pay the arbitration costs retrospectively, because he has wasted the State's time and money. He may also have damaged the Warrant-holder by identifying a gap in his protection which would be open to exploitation by a third party. The knowledge of this possible penalty would deter any potential investor who was not completely serious about his project.

Professor Silberston makes the point that it is odd to consider making legal aid available in respect of direct protection of innovation, when it does not exist for classical Patents.

The new legal arrangements proposed do also appear to have great potential for other forms of intellectual property. In the case of classical Patents, the large numbers are intimidating, but this problem could be dealt with by restricting eligibility to *exploited* Patents. The criterion would then be substantially the same as that for both direct protection systems, and, as in their case, the sums of money involved could be expected to be quite small.

It must be obvious that even if Direct Protection of Innovation did not have many other advantages, its arbitration provisions alone would remove what is perhaps the most serious deterrent of all to investment in innovation, especially by small- and medium-sized firms. The strength of the resulting stimulus to innovatory activity can only be imagined.

CHANGES IN PROPOSALS

As a result of the experts' comments, and reflection on these, it has been possible to improve the proposals advanced in Chapters II and III in several respects. These changes may be summarised as follows:-

1. The individual "variable term" arrangements of both Innovation Patent and Innovation Warrant will be replaced, at least initially, by three fixed, "project-related" terms.
2. The Innovation Warrant's principle that the State should enforce the monopoly grants it makes, will be expressed in the arrangements for arbitration and legal aid, outlined above.
3. The latter arrangements will also apply to the Innovation Patent.
4. The Innovation Warrant will now adopt the Innovation Patent's proposals for dealing with possible conflict with classical Patents.

It is submitted that Direct Protection of Innovation, as now formulated, offers to the countries of the E.E.C. a most valuable opportunity of making successful industrial innovation commonplace, instead of exceptional. It offers a counterbalance to the strength of the United States in goods that depend heavily on modern marketing techniques, and an alternative to protectionism in dealing with competition from Japan. It could contribute to employment and also to achievement of the better economic balance between the Regions that is one of the objectives of the Rome Treaty.

NEED FOR PRACTICAL EXPERIENCE

When the Canadian Working Party on the Patent System reported in 1976, it proposed that a new Patent Act should be passed immediately, which would explicitly have a duration of only ten years. This Act would correct some of the features of the classical Patent system which had been identified by the Working Party as clearly irreconcilable with Canada's economic interest. The ten-year period would then be used to investigate other doubtful features, so that a soundly-based decision could eventually be made as to whether Canada should have a Patent system at all, and if so, what its structure should be.

PROPOSALS FOR PILOT SCHEME(S)

A similarly "investigative" approach would be appropriate in the present case. The Commission of the European Communities is giving a high priority to innovation, and has a substantial budget for supporting projects to this end. It is recommended that part of these funds should be directed towards support of pilot schemes of Direct Innovation Protection in some Member-States, with a view to developing Community-wide arrangements as soon as possible, on the basis of actual experience of working the system. Even if they have to overcome difficulties due to inexperience, any member States which establish a pilot scheme, could expect to gain substantially from it.

SOCIAL INNOVATION

The main reason why such pilot schemes are needed, is that social innovation can be no different from technological innovation in terms of the need for "learning by doing". There are of course several aspects of the proposals discussed in this Report which could advantageously be investigated further, in the academic sense. However, no abstract studies can produce the essential information on the actual performance of the ideas in practice, which would be needed to enable decisions on a Community-wide system to be taken on a sound basis.

If the Commission decided to accept this recommendation, it would presumably do so by offering to match resources made available by individual member-States for schemes of direct protection of innovation of limited duration. It would be open to any member-State to take the initiative in its own case. The value to the Community as a whole in terms of information generated from experience, would obviously be greater if one of the larger countries, such as France, were to operate such a pilot scheme. The work-load of the Netherlands Patent Office has declined greatly since the establishment of the European Patent Office. This might mean that there is available there a cadre of experienced and qualified personnel, which could be immediately employed in operating the new system. It is also worth recalling that a vacuum is created in Spain's Intellectual Property arrangements through the suppression of the local Patent of Importation on joining the Community, and a pilot scheme of direct innovation protection might find a corresponding place there.

It goes without saying that any such pilot scheme should leave the existing Patent system untouched, except for the minor changes mentioned above to provide for any case of conflict between it and the arrangements for direct protection of innovation. Both Innovation Patents and Innovation Warrants should be offered in any pilot scheme. Each of these has special features which the other lacks, and nothing except actual operating experience can be a reliable guide to the relative value of these features in the eyes of innovators, investors and administrators.

During the trial period, therefore, it should be possible to use either of the two approaches to direct protection of innovation, according to the circumstances under which a new project is being undertaken. Many business men and investors could be expected to prefer to know the exact limits of their monopoly before making an investment, and this is offered to them by the Innovation Warrant; others might see more advantage in maximising their lead time. Since with an Innovation Patent the first that competitors would know of the existence of a new product would be when it arrives on the market, all evaluation by the Innovation Office taking place after that date, and all decisions being retrospective to it, this is then likely to be the preferred type of protection.

SPECIAL ADVANTAGES

Also, each type of direct protection of innovation has special advantages to offer different sizes of firm. For example, Professor Silberston is quite correct to point out that it is the medium-sized firm that has most to gain from the Innovation Warrant proposals. But against this, these proposals have no features that could offer the same attractions to the private inventor as the Innovation Patent. This is only one example of how a protection system which offered the features of both Innovation Patents and Innovation Warrants would cover the spectrum of firm size better than one which offered only one of the proposed sets of monopoly arrangements.

Facilities for arbitration would have to be established in any event for the Innovation Warrant, and these could be used for whatever cases of infringement (presumably infrequent and relating to "option" claims only) might arise out of Innovation Patent disputes. Since the novelty criterion is the same for each type of direct protection, the same machinery for assessing it would serve for both.

As to the variable term of monopoly, any pilot operation could commence with the three-term system advocated by Professor de Jong. It would be essential, however, to establish research capacity from the start, both to learn how to match terms as precisely as possible to the need for them, and to investigate the possibilities of the Innovation Patent's special provisions for licensing. The Research Group in each national Innovation Office would have the additional duties of measuring all aspects of its pilot scheme, and of liaison with those responsible for monitoring progress and possibly developing Community-wide proposals, for the Commission.

STRUCTURE OF TYPICAL PILOT SCHEME

It is now possible to see the outlines of a set of practical arrangements for direct protection of innovation which could be used in any pilot scheme. These take account of present realities, and also have the potential for future development. The following features would be included:–

1. National monopolies would be granted so as to benefit those who perform innovatory economic acts.
2. These monopolies would be granted and administered by a publicly controlled Innovation Office, separate from the Patent Office.
3. "Innovatory economic acts" are defined as actually putting on the market, products which have hitherto not been available in the ordinary course of trade within the European Community, or which have not been so available as the result of a particular process of manufacture.
4. When products without a particular feature have already been available in the ordinary course of trade, the monopoly will only relate to products

which incorporate that feature. Similarly, if products were available earlier as a result of another process of manufacture, the monopoly will apply only to the new process.

5. After making its own enquiries, and before making any monopoly grant, the Innovation Office will give all interested parties appropriate opportunities to object.

6. Any grant of monopoly will only be contestable on the ground that it had been obtained by fraud.

7. The Innovation Office will make no charge for dealing with applications, nor for keeping grants in force.

8. Every grant will be for one of three terms. The longest term will be for radical innovation, and the shortest, for incremental innovation.

9. The Innovation Office will be charged with the duty, and provided with the resources, to make its monopoly terms correspond progressively more closely with what is needed to reward innovators for the risks they run.

10. The Innovation Office will have power to modify the number or length of the monopoly terms it grants at its own discretion, without the need for further legislation.

11. In all cases of infringement or conflict, the Innovation Office will recruit from the business or scientific communities, preferably with the co-operation of the parties, an individual who is expert in the precise subject-matter that is in dispute. Each party will state its case in writing, and after study of the cases, the Arbitrator will hold one or more oral hearings, so as to clarify the issues to his own satisfaction. No legal representation will be permitted at a hearing. Witnesses who could contribute to the case, may be suggested to the Arbitrator by either party, but only the Arbitrator will have the right to call witnesses, expert or otherwise.

12. Any decision of an Arbitrator may be appealed to the Courts.

13. Both parties will be required to accept, and if necessary to act upon, an Arbitrator's decision without delay. Failure to do so, or to co-operate with the Arbitrator in every way, will result in the Innovation Office placing resources at the disposal of the other party, to pursue the matter in the Courts.

14. The amount of this legal aid will be at least what is required to match any resources deployed in litigation on its own account by the opposing party.

15. The Innovation Office will not charge for its Arbitration services, but any person or firm who is responsible for an unnecessary Arbitration procedure will have to bear the cost of it.

16. An application for a monopoly will specify that it is to be in the form either of an Innovation Patent or of an Innovation Warrant.

17. The Innovation Office will draw up rules of procedure in accordance with Chapter II above, for the grant of Innovation Patents, and in accordance with Chapter III above, for the grant of Innovation Warrants, and in accordance with points 8-15 herein, for both.
18. The first monopolies to be granted by an Innovation Office will be limited to Engineering products and processes. The Office will have power to include additional fields at its own discretion without any need for further legislation.
19. Similarly, the first monopolies in any pilot scheme will be available only to applicants located in an area which qualifies for specific regional aid. The Office will have power to include additional areas at its own discretion without any need for further legislation.
20. The funding of any pilot scheme will represent an alternative use for public money allocated for the encouragement of innovation.
21. Monopolies granted under a pilot scheme may outlive the scheme itself; since the latter may be replaced by Community-wide arrangements.

FINAL SUMMARY

The case for Direct Protection of Innovation has been strengthened by being placed in the context of the growing information economy, and the proposals for carrying it out in practice have been improved significantly.

Arbitration and Legal Aid, as now proposed, constitute a major advance, and also have great potential for other types of intellectual property.

The Commission of the European Communities is recommended to support operation of Direct Protection of Innovation in member countries on a pilot basis.

This development could produce benefits for the public of the same order of magnitude as legislation for Limited Liability did in its day. That legislation massively reduced the perceived risks in investing in industrial development, which subsequently grew correspondingly.

Direct protection of innovation could have similar results by reducing the level of risk involved in investment in innovation, especially as this is perceived by investors.

NOTES

1. T.S. Kuhn *in* R.R. Nelson (ed.): The Rate and Direction of Inventive Activity, Princeton Univ. Press. 1962, p.453; Whitehead, A.N.: Science and the Modern World, Cambridge 1947, p. 14.

2. Passer, H.C. The Electrical Manufacturers 1875–1900. Univ. of Chicago Press, 1969.
3. Nelson, R.R. and Langlois, R.N.: Industrial Innovation Policy: Lessons from American History. *Science*, 18 Feb. 1983.
4. Strassmann, W.P.: Risk and Technological Innovation. Cornell Univ. Press, 1969, p 79.
5. Nordhaus, W.A.: Invention, Growth and Welfare. M.I.T. Press, 1969.
6. Edwin Mansfield, Mark Schwartz and Samuel Wagner: Imitation Costs and Patents: An Empirical Study. *Economic Journal* December 1981, pp. 907 – 919.
7. Report of American Research and Development Corporation, 1966.
8. Baran, P.A. and Sweezy, P.M.: Monopoly Capital, London 1966.
9. Scherer, Frederic M.: The World Productivity Growth Slump. IIM/IP Conference paper, 1984. Wissenschaftszentrum, Berlin.
10. Evenson, Robert E.: International Invention: Implications for Technology Market Analysis. *In* Zvi Grilliches (ed.): R&D, Patents and Productivity. Univ. of Chicago Press 1984. pp. 99–122.
11. Mansfield et al., op.cit.
12. Levin, Klevorick, Nelson & Winter: Survey Research on R&D Appropriability and Technological Opportunity. Part I: Appropriability. Yale University Working Paper, July, 1984.
13. Nordhaus, op.cit. p.56.
14. Kingston, W.: Innovation. London 1977, Ch.II.
15. Pendleton, Michael: Intellectual Property, Information-based society and a new international economic order – the policy options. *European Intellectual Property Review* [1985] 2, p.31.
16. Brown, Donald L.: *Journal of the Patent Office Society* July 1960, pp. 439–455.
17. Tullock, Gordon: Trials on Trial, Columbia Univ. Press 1980.
18. ibid. p.118.
19. ibid. p.113.

342

344